HONEY BEE ALCHEMY

A contemporary look at the mysterious world of bees, hive products and health

Valery A. Isidorov

IBRA
INTERNATIONAL BEE RESEARCH ASSOCIATION
NBB

Honey Bee Alchemy
A contemporary look at the mysterious world of bees, hive products and health

This edition is jointly published by and available from either:

International Bee Research Association at www.ibra.org.uk
Tel (0044) (01) 1769 572401

or

Northern Bee Books at www.northernbeebooks.co.uk
Tel (0044) (01) 1422 882751

ISBN: 978-1-913811-02-0

Design and artwork by DM Design & Print

The author of the book VALERY ISIDOROV is a professor, full member of the Russian Academy of Natural Sciences. He is a graduate of the Faculty of Chemistry of the Leningrad (now St. Petersburg) University. His scientific interests lie in the field of environmental chemistry and the chemistry of natural products. He is the author of over 150 scientific publications and 12 monographs published in English, Russian and Polish. For more than 20 years he worked at the Institute of Chemistry of the University of Bialystok, where he created and for many years headed the Department of Environmental Chemistry. Currently, Professor Isidorov runs a laboratory at the Institute of Forest Sciences of the Bialystok Technical University.

Contents

Preface to the English edition 6
Foreword 7
Chapter 1. Introduction 11
Chapter 2. "Information technology" of a honey bee colony 19
2.1. Information is the mother of the order 19
2.2. Chemical "levers of power" of a queen 25
2.3. Info-chemistry of worker bees 32
2.4. Pheromones of larvae give a signal: We are here, feed us! 37
2.5. The drones "Men's club" 38
2.6. Pheromones at the service of beekeepers 40
Chapter 3. The mysterious royal jelly 43
3.1. Myths and secrets of royal jelly 43
3.2. Fateful food: The larva's diet determines future fate of a bee 44
3.3. Royal jelly – a panacea for all diseases? 49
3.4. Not only a "food warehouse" but also a real "arsenal" 54
3.4.1. A volatile defensive weapon in royal jelly: against whom it is directed? 55
3.4.2. Why is royal jelly "too tough" for microbes? 58
3.5. Useful "impurities" of royal jelly. 70
Chapter 4. Drone homogenate: invaluable product of beekeeping 73
Chapter 5. Propolis: ***urbi et orbi*** 81
5.1. Propolis is more than glue 81
5.2. Work of bees to collect herbal balms: workaholics and lazy people 83
5.3. Is it possible to establish the origin of propolis? 90
5.4. What does propolis smell like? 109
5.5. What does propolis treat and how? 116
Chapter 6. What's new about honey? 132
6.1. Is it worth continuing to research it? 132
6.2. Sweet contribution of plants paid to bees 133
6.3. The way from nectar to honey – the order of magical transformations 139

6.4. Unifloral honeys and their plant precursors 148
6.5. A new category of beekeeping products: Medical-grade honey 155
Chapter 7. Not from nectar or honeydew, but still honey 160
7.1. Required explanation 160
7.2. Honey or not honey? That's the question 161
7.3. Herbhoneys differ from honey but not for the worse 164
Chapter 8. Beebread - Food of the Gods 174
8.1. Bee pollen and beebread, are they one and the same? 174
8.2. No magic, just an ordinary chemical miracle 176
8.3. Beebread through the prism of chemical analysis 179
Chapter 9. Stay healthy, bees! 185
9.1. About the health and treatment of bees in general terms 185
9.2. The most endangered species 188
9.3. Varroatosis 191
9.3.1. Anamnesis and current condition of the patient 191
9.3.2. "Case history" and the patient's condition 192
9.3.3. Current approaches to combat varroatosis 196
9.3.4. Essential oils and other natural remedies 204
9 3.5. Biological methods of dealing with varroatosis 209
9.3.6. Zoo technical approaches 217
9.3.7. Medication-free treatment 220
9.4. Bee quarantine disease: American foulbrood 226
9.4.1. The causative agent and the clinical picture of the disease 226
9.4.2. Bacteria against AFB 229
9.4.3. Natural remedies for American foulbrood 234
9.5. Infestations of pests and pathogens continue 237
References 239

Preface to the English edition

The book offered to the reader's attention is the author's translation of a book published in Poland in 2013. The book is addressed to the lovers of wildlife and honey bees, and to everyone interested in beekeeping products, their composition and medicinal properties. For Western European and American readers, Chapters 4 and 7, containing information about practically unknown in the West unconventional but very valuable beekeeping products, may be of particular interest.

The author's scientific interests extend not only to the chemical composition and the origin of propolis, honey and other bee products, but also to the bee health. Unfortunately, in recent decades the bee health has noticeably deteriorated, resulting in many difficulties in beekeeping. The last Chapter of the book is devoted to the bee health and possible ways of their treatment, avoiding synthetic drugs with their unwanted side effects. The author's work in recent years is connected with these problems and it is no coincidence that the chapter devoted to the bee health has turned out to be the most revised and extended in comparison to the Polish edition.

In preparing this publication, photographs were used, kindly provided to me by the Polish beekeeper Mrs. Teresa Porankiewicz-Bartkowiak, to whom I express my sincere gratitude.

Foreword

"Thus you bees make honey, but not only for yourselves"
Virgil

For thousands of years, man has used honey, bee wax and bee venom as well as propolis. Not long ago, other bee products, such as royal jelly, bee bread and bee pollen (pollen load), were also used, mainly in medicine, cosmetology and for dietary purposes. However, almost until the second half of the twentieth century, we had only a vague idea about the chemical composition of these products and about those substances that form their useful properties, namely taste and healing.

People knew long ago that honey helps to preserve products from spoiling, but not long ago they knew nothing about the components that are responsible for this property. Sugar? Of course. We all know pretty well that strong sugar syrup, and jams made from it, can be stored in the cold for a long time without being spoiled. Nevertheless, not so long ago, honey and honey processed foods, including fresh meat, were not stored in the refrigerator. This indicates that the reason for this is not only the high content of sugar: there is something in honey that kills microbes, or at least does not give them the opportunity to expand. But even now we can read in some research works that antimicrobial honey properties are connected with a small content of admixture, "...which nature has not yet investigated".

Honey. We know that it consists of 82–85 % glucose and fructose and 15–18 % water. And there is also something that gives it a unique aroma. And what about propolis? At least 50–60 years ago, a question about composition of propolis would stump anyone, though it is known that propolis has been used in medicine since antiquity.

So perhaps one should not scratch one's head over such questions? It works and thank God for that! But first, the desire to learn about the world around us is inherent to man – there is nothing one enjoys better than the knowledge of something new. Secondly, new knowledge about the nature of something lets us use it consciously and for our good. It is important that this knowledge is not used to struggle against nature, which borders on an act of violence against it.

Huge progress in the development of chemical analysis technology has given us vast opportunities to research different kinds of products that differ in their complex compositions, including those produced by bees. The leading role in such research belongs to chromatography. The author of this book has devoted many years to researching the content of bee products and their botanical precursors with the help of one of the most effective varieties of this method – gas chromatography combined with mass spectrometry (GC-MS).

In order to understand the so-called "chemistry" of honey, propolis, bee bread and royal jelly, I had to come into direct contact with their creators – honey bees. I had to remember the days of my youth when I watched attentively and helped my grandfather with his small apiary. I also had to read vast numbers of books and papers devoted to the chemical composition of these products, bee biology and bee behaviour. While reading scientific literature about bee products, I was struck by the huge disproportion in the number of publications devoted to the research of each of these products. Below is a diagram showing the number of articles in all the scientific magazines produced by the two largest publishers – Elsevier and Springer (they issue hundreds of such magazines) – that have been published within the last 10 years. As we can see, the greatest attention of researchers is given to honey (almost 13,000 articles!), while all aspects concerning propolis – its composition, origin and use in medicine – are covered less often. Even fewer publications are devoted to royal jelly and bee wax, and just a few of them to bee bread.

That is why, first and foremost, I and my colleagues began to study the chemical composition of this valuable product. Very soon the range of our interests expanded and covered other bee products, namely honey, propolis and royal jelly, whose content can be studied using the GC-MS method.

I should confess that in more than 45 years of research I have never experienced more pleasure than in the course of these investigations. Well, probably with the exception of the years of study devoted to volatile organic substances, i.e. those released into the atmosphere by living plants. The richness and singularity of identical products, as they may seem at first, whether it is propolis gathered from a nearby beehive or monofloral varieties of honey from a nearby apiary, makes their study a fascinating experience. Incredible satisfaction is brought about by the search for, and discovery of, general regularities in the formation of their composition, and the discovery of new unknown compounds. To identify the similarities and peculiarities of different bee products' chemical composition, we had to expand their "geography" and analyse many hundreds of tests from different countries of the Eurasian continent and different geographical areas.

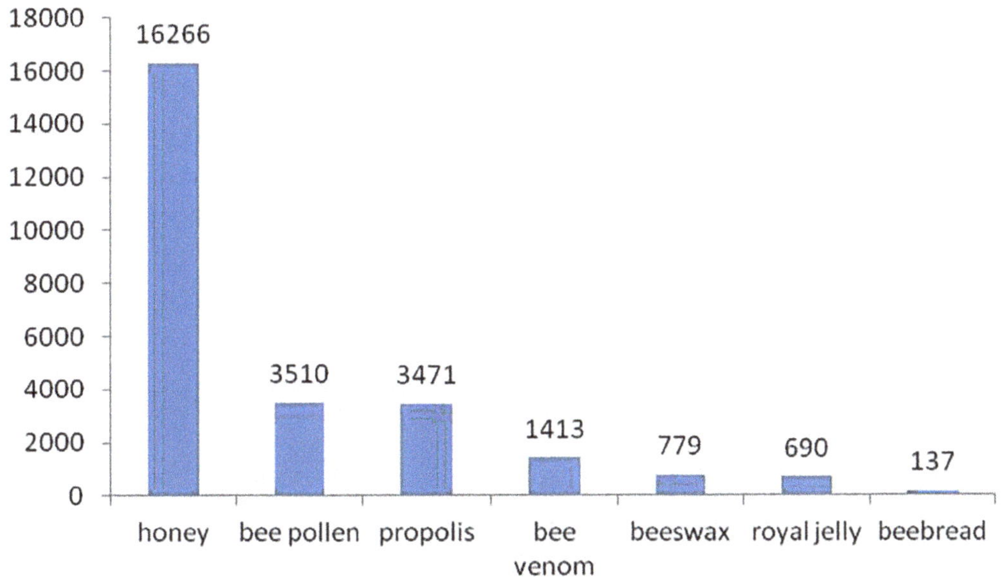

Figure 1. The number of scientific works devoted to different kinds of bee products published within the last 10 years by the publishing companies Elsevier and Springer.

The global literature devoted to the theme of bees and their products includes not only huge numbers of publications in scientific magazines but also dozens of scientific and popular books. However, more and more new books are continuing to appear and the reader holds one of them in their hands. In setting about working on writing the manuscript of the book I have in front of me, to be frank, it is a difficult task to introduce its subject in such a way as to satisfy the interest of as wide a range of readership as possible. There are those who are merely interested in information about the life of bees and bee products and those who deal with the subject in a professional way. In order not to make reading the book too difficult, but at the same time to give more complete results from my work, as well as that of the other researchers I referred to earlier, I tried to minimize the experimental data, and the number of tables and diagrams. In the literature cited at the end of the book, the interested reader can find additional information about the biology of bees, about the medicinal properties and chemical composition of honey, propolis, royal jelly, bee bread, as well as their botanical precursors – about nectar and pollen of some species of honey plants, and also balsamic secretions that cover the buds of some tree species. I hope this data will be useful to those who have embarked on a difficult, but exciting and fruitful way of studying the composition of beekeeping products, as well as other natural objects. As for the results of my research and whether it corresponds to a set goal – you are the judge, my reader.

I consider it my pleasant duty to express my sincere gratitude to my colleagues and students of different years who, to one degree or another, took part in the collection of the experimental data presented in this book. In conclusion, I would like to thank my wife Vera Vinogorova for her daily support, patience and help in preparing the manuscript.

Chapter 1
Introduction

For at least 5,000 years, the fate of man has been connected with that of the honey bee, and during this time so many words of praise have been declared, sung and devoted to these little workers that it might seem impossible to add anything else, even if one wanted to. Great minds, beginning with Plato and Aristotle, and great poets, including Virgil, Ovid and Shakespeare, not only drew inspiration for their work from, but also did not forget to sing their praises to, bees. In 1969, the world-famous bee biology researcher Karl von Frisch observed: “Out of all animals living on the Earth, the most favoured is probably the bee.”

Honey bees have always not only been favoured but also highly respected. An obvious example is the use of bee images in heraldry. The privilege of using the image of bees was enjoyed by those cities whose goods were known all over the world, and their citizens were distinguished by hard work.

Figure 1.1. Coats of Arms of Manchester City Council (1842), and Russian cities Tambov (1730) and Medyń (1777).

The concern over the future of honey bees and their ability to provide their products is reflected in the symbol of the first world savings bank in the capital of Austria, Vienna.

Figure 1.2. Pediment of the first savings bank on Graben Street in Vienna.

Bees' unceasing industriousness, wonderful constructive art, altruism and allegiance to the cause – all these traits inspire the admiration of not only philosophers and poets but many statesmen as well. According to some sources, the legendary legislator Lycurgus, while creating the social and political system of Sparta, took as a model a honey bee family with its perfect order. It is easy to believe this legend: not only in high antiquity but also in recent times the organization of this family seems to be exemplary. Here are the words of the Archbishop of Canterbury in one of Shakespeare's plays, referring to the young king and nobility:

...for so work the honey-bees,
Creatures that by a rule in nature teach
The act of order to a peopled kingdom.
They have a king and officers of sorts;
Where some, like magistrates, correct at home,
Others, like merchants, venture trade abroad,
Others, like soldiers, armed in their stings,
Make boot upon the summer's velvet buds,
Which pillage they with merry march bring home
To the tent-royal of their emperor;
Who, busied in his majesty, surveys
The singing masons building roofs of gold,
The civil citizens kneading up the honey,
The poor mechanic porters crowding in
Their heavy burdens at his narrow gate,
The sad-eyed justice, with his surly hum,

Delivering o'er to executors pale
The lazy yawning drone.

William Shakespeare, *King Henry the Fifth,*
Act I, Scene 2

In these verses one can see the ideas about the life of bees that were so typical at the end of the sixteenth century. I should mention that they had not changed much since the days of Aristotle, who summarized the accumulated knowledge about honey bees over the preceding 2,500 years. According to Aristotle, a "king" ruled a bee family; and all that could be said about drones was that they are "larger than others, have no sting and ...stupid". Aristotle distinguished drones as being a special caste which "...harms other bees".

Honey bees do not display their life for all to see, which is why even such a shrewd scientist as Aristotle remained in the dark about many aspects of their life, in particular the most "intimate" matters concerning reproduction and breeding. Readers today, even if not particularly well versed in such matters, cannot help smiling at reading such lines as:

We know full well
that if you take a sacrificial bull
and bury the tossed-out carcass in a ditch,
from every portion of the putrid entrails
flower-sipping bees will rise. And these bees,
just like the animal from which they spring,
live in the fields, love toil, and work with hope.
Ovid, Metamorphoses, Book 15, lines 364–365

In spite of its evident absurdity, this "belief" survived until the seventeenth century and was included in beekeeping books. It is terrible to think how many bulls, as sacrificial "parents" fell victims to naïve readers! Ovid himself did not doubt for a moment the truth of this breeding method of the honey bee . He introduced the fragment cited above with the words: "But we should trust phenomena that are fully proved".

The end to the ignorance that had prevailed for centuries regarding bees' breeding and reproduction came at the turn of the sixteenth and seventeenth centuries: in 1586 a Spanish researcher, Luis Mendez de Torres, published his work about a queen that laid eggs, and half a century later Charles Butler (Butler, 2017)[1] published: *The Feminine Monarchie or the Histori of Bees. Written out of Experience by Charls Butler, Magd.*

This book, the content of which was based on the author's personal observations, not only talked about the reign of a "queen" in a hive but also about the role of drones in bee reproduction: he stated that "droneless hives become infertile", and hence in a family without drones breeding did not take place.

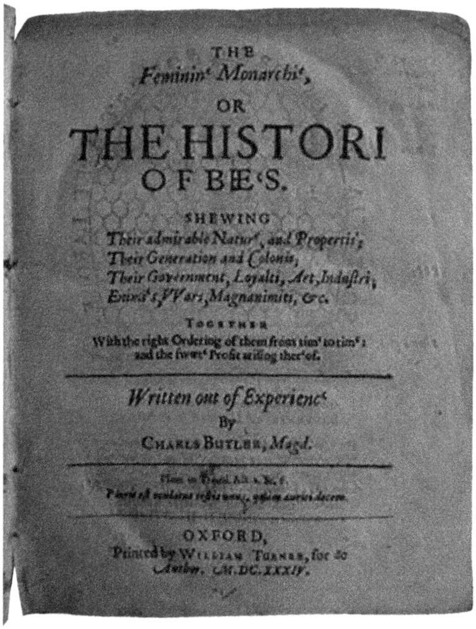
THE
Feminin' Monarchi',
OR
THE HISTORI
OF BEE'S.
SHEWING
Their admirable Natur', and Propertis;
Their Generation and Colonis,
Their Government, Loyalti, Art, Industri,
Enemi's, VVars, Magnanimiti, &c.
TOGETHER
With the right Ordering of them from tim' to tim':
and the sweet' Profit arising ther'of.

Written out of Experienc'
By
CHARLS BUTLER, Magd.

OXFORD,
Printed by WILLIAM TURNER, for the Author. M.DC.XXXIV.

Figure 1.3. The Feminine Monarchie, originally published by Joseph Barnes, Oxford, in 1609

The Latin binomial, generic and specific name of the honey bee, *Apis mellifera* was given in the middle of the 18th century by the creator of the metaphysical taxonomy of animals and plants, Karl Linnaeus. It is probably no coincidence that images of Linnaeus (obverse) and a bee collecting nectar (reverse) are shown on the Swedish 100 krona banknote.

---

1 *Footnote: Reverent Butler (he was a vicar in Hampshire) took interest not only in bees but was also an active follower of English orthography reforms and its simplification. This was embodied in the title page of his book: Monarchie instead Monarchy, Histori instead of History and Charls instead of Charles.*

Figure 1.4. Swedish 100 krona banknote depicting Linnaeus and a bee collecting nectar

Further progress in the understanding of the reproduction biology of a honey bee is associated with the names of researchers from different countries (Crane, 1999). Among them are – the Dutchman Jan Swammerdam, who dotted the "i's" regarding the question of the queen's gender, a worker and a drone. In the 18th century, the Lusatian Adam Gottlieb Schirach (he was a Sorbian Protestant pastor and permanent secretary of the physical-economic bee society in Upper Lusatia.) associated his practical and journalistic activity with Silesia to a large extent (Tomazewski, 2011). He determined that a queen and a worker develop from similar eggs and that in a queenless family workers build up cells with worker larvae (emergency queen-cell is lid) and breed queens.

Slovene Anton Janša, who was taken into Austrian empress Maria-Teresia's service as a teacher in a beekeeping school, noticed that bees in queenless families became laying workers ("anarchistic queens"), that a non-fertilized queen lay only drone eggs, and the first swarm leaves a hive together with a queen. It was he who, in 1771, first described the mating flight of a queen her insemination away from the hive. Later, in 1788 this fact was corroborated by Swiss scientist François Huber who is considered to be one of the founders of the science of bees. A huge contribution to this science was made by Polish scientist Jan Dzierżon (Tomaszewski, 2011), who, in 1835, discovered the parthenogenesis of honey bees (described for the first time in a publication in 1845 and finally confirmed in 1853). He was considered to be the most competent bee scientist and practitioner of his time. It is remarkable that he is the only researcher of bee life after whom a city has been named (Dierżoniów, Poland).

Jan Swammerdam (1637–1680) *Adam G. Schirach (1723–1783)* *Anton Janša (1734–1773)*

François Huber (1750-1831) *Jan Dzierżon (1811–1906)*

Figure 1.5. Scientists and beekeepers who have made significant contributions to the study of the biology of honey bees

As we can see, secret elements of bee reproduction and breeding have been relatively recently discovered. It is remarkable that new ideas, even when proved by careful observations and experiments, were accepted with either with incredulity (as was the case with Schirach) or with furious resistance by the scientific "establishment" (as Dzierżon had to face). Further progress in the research of bee life was made in a detailed study of their anatomy (Russian scientists N. Nasonov and G. Kozhevnikov contributed to this field) and mainly with the development of scientific selection based on knowledge of genetic heredity law.

N.V. Nasonov (1855–1939) *G.A. Koschevnikov (1866–1933)*

Figure 1.6. Russian biologists, researchers of the anatomy of honey bees, whose names are named after the glands they discovered.

The study of the honey bee's chemical communication and the chemical composition of their products started much later, at the beginning of the twentieth century. However, real development took place in the second half of the twentieth century when many new methods of research appeared, such as spectroscopy, chromatography and mass spectrometry. Subsequent chapters of the present book are devoted to the discoveries that have been achieved within the last decades due to such methods, and the secrets of bee alchemy that still remain undiscovered.

Chapter 2
"Information technology" of a honey bee colony

2.1. Information is the mother of order

Honey bees (*Apis mellifera* L.) belong to the group of social insects. Moreover, some authors (Tarpy & Gilley, 2004) consider them to be *truly* social (eusocial) insects. Eusociality is characterized first of all by the collective and devoted care of members of the colony of the brood. Secondly, the bees that cannot provide brood, work collectively for the good of their fertile congeners and help them in every way. Due to this collectivity and compatibility, the honey bee colony resembles in many ways an integral multicellular organism because of the coordination of the behaviour of all its "special organs" performing definite functions. Some scientists use the special term "superorganism"[2] (Seeley, 1989; Moritz & Fuchs, 1998; Tautz, 2008). What parts does this organism contain? What are their functions? And how is the coordination of their activity organized?

It is known that a honey bee colony contains the queen, some tens of thousands of worker bees and a considerably smaller number of drones. Brood production is performed by the queen and drones (they are sort of the reproductive organs of the colony): all other work is done by worker bees of the same gender as the queen. Queens lay fertile or infertile eggs in the cells prepared by worker bees and this determines the gender of future bees: female insects hatch from fertile eggs, drones – from infertile eggs.

So in the honey bee colony, female bees predominate while the ovaries and other generative organs of worker bees are not fully developed and as a result they are not perfect – they are "functionally sterile", as the specialists say. Nature gave them all the necessary working "tools" instead: a long proboscis for acquisition of nectar and water; pollen baskets on the hind legs to gather pollen; and strong mandibles for

---

2 *This term was introduced in 1911 by American biologist W.M. Wheeler. It should be noticed, that the concept of superorganism has its critics who consider that the similarity between highly organized insects and multicellular organism is strained. Some arguments of the critics are discussed in Moritz & Fuchs's work. (1998).*

chewing pollen, building combs and collecting sticky substances used for making propolis.

Worker bees are likely to have a lot of different duties to maintain order and safety in the colony. Aside from those listed above, their functions are:

- preparing (cleaning and polishing) the cells for the queen's laying of eggs;
- taking care of the queen and brood (feeding and heating of larvae, closing the cells using the wax cappings);
- converting nectar, preparing honey and bee bread for long storage (capping the cells with honey and closing the cells with bee bread using honey);
- maintaining the nest.

Surveys show that this kind of work is done by all worker bees. When a new bee emerges from a cell, it will first prepare its cell, and then it becomes a "nurse bee" for larvae, the queen and young drones, because these "egoists" cannot feed themselves for the first time. Later, all young, not flying bees begin to receive and convert the natural blessings brought by the oldest members of the family – the forager bees.

Besides all these "fundamental" tasks, there are some more rare ones. Some of the bees (5–10%) become "defenders", guarding the nest from robber bees and other aliens. A smaller number of worker bees do the dirty work as "aid-men" and as members of the "funeral detachment": they take away dead bees and other stinking things from the nest. Probably they also perform "police" functions: watching the pseudo-queens and replacing the eggs laid by them. And, at last, some bees that are the same age as forager bees but do not perform the function of foraging become "defenders", guarding the external part of the nest.

You can imagine that there might be chaos in a big colony that numbers tens of thousands (up to 80 thousand) bees performing a great number of functions! But that is not the case: in the hive "let every herring hang by its own gill"; in other words, each must take responsibility for their own actions. This is due to the general system of the distribution of responsibilities, which is called "polyethism". The main principle is the division of labour with different members of the family doing different work.

***Polyethism** – is a fixed difference in carrying out definite functions by social insects. Species with the same physiological features and disposed to do the same work compose polyethical groups. There is caste polyethism and also age polyethism. The term "polyethism" was first employed by J.S. Weir in 1958.*

Caste polyethism is typical of many species of ants when division of different castes is based on difference in anatomical characteristics of species. For example, ants from the "soldiers" caste are bigger and have strong mandibles that allow them to bite through the thief ant and other enemies. Such a "soldier" guards all its life. Worker ants also differ by size and have a special anatomical construction to carry out definite functions.

The situation with the honey bee is different and the evolution process of these social insects – ants and bees – differs. All worker bees in one family are sisters and if they differ in anatomical construction, this difference is not so significant. The division of labour among them – changing one polyethical group to the other – occurs over time, with the increase in age. Age polyethism of worker bees consists of changing the type of work, which is closely connected with the development of some glands and their progressive atrophy. Newborn worker bees have inactive mandibular and hypopharyngeal glands as well as wax glands. The sting apparatus is not yet developed (in comparison to the queen, who meets a situation head-on). That is why the function of young worker bees is cleaning and polishing cells. A worker bee becomes a nurse bee with the development of the mandibular and pharyngeal glands (after the third day of life), which produced royal jelly. On the 9th –12th day of life, these glands become fully developed. At the age of 12–18 days, the wax glands become fully developed (although at the age of 4 days, the worker bee may be given the opportunity to take part in building), but by the end of the third week the function of the wax glands becomes weaker and then they stop emitting wax. By this time the mandibular glands are shrunk and a bee's pharyngeal glands that are older than 20 days produce a new substance, namely invertase, which is an enzyme that catalyzes the hydrolysis (breakdown) of sucrose into the monosaccharides fructose and glucose. Now a bee is ready for foraging. This enzyme is also necessary for forager bees collecting nectar. As you know, foraging occurs in the final stage of a bee's life.

This is the *general course* of a bee changing from one polyethical group to another. But as previously mentioned, there are other bees who have a "rare profession" that do not belong to this process. How is the recruiting done? Why are some bees made to do the dirty work, such as "aid-man" or "coffin maker"? Why don't some forager bees continue their work in their last days of life but only guard the nest? Also there are some differences in the behaviour of bees following the general course: some forager bees prefer collecting nectar, some prefer bringing pollen and only a few of them bring both substances. Is this accidental? It is time to remember that although all worker bees in the colony are sisters, they are not "sisters to the full extent", because a young queen mates with up to 25 drones (and this is not the limit in the world of bees: in the colonies of a big Indian bee *A. dorsata* one queen can mate with as many as 35). As a result, the colony contains a lot of subfamilies that differ in their combination of genes.

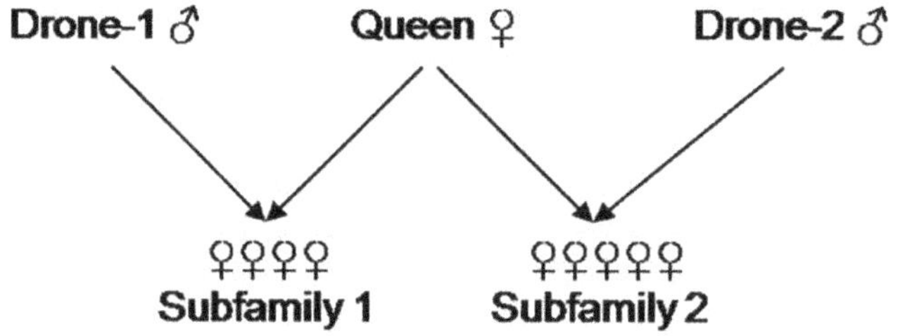

Figure 2.1. Although all worker bees in the colony are sisters, not all of them are "sisters in full measure", because they come from different fathers.

You can believe that polyandry, and as a result the "genetic diversity" of bee families, is an important factor in their ecology. It was found out that in a family where the queen is fertilized artificially by the semen of one drone, the degree of diseases such as European and American foulbrood and many others is higher than in those families where the queens are fertilized by the semen of ten different drones (Tarpy & Seeley, 2006).

Scientists have drawn attention to the interesting consequence of the non-homogeneity of a bee colony – i.e. nepotism (this phenomenon is typical of human society). The protection of full sisters is noticed, for example, in different periods of breeding new queens, from larvae to the virgin queen bee that is ready to emerge (Tarpy & Gilley, 2004). But in comparison to human society, nepotism in bee colonies is not the most evident and "half-sisters" are not strongly discriminated against.

That is why full sisters (the brood of the queen and the same drone) and half-sisters (the brood of the queen and other drones), with all their similarities, differ by some features and, as the scientists Calderone and Page, Jr. (1988) suggests, why they have a genetic predisposition to special types of work. It may that an inherited olfactory super-sensitivity makes some bees clear up the dead bodies and other smelly things from the nest because they don't like the smell of decay. The future "guards" also have to recognize members of the family by smell, even if they are only "half-sisters" and their smell isn't the same as that of "full sisters". What makes some bees become "soldiers" and guard the outside part of the nest for their entire life? Congenital anger? Genotypic (inherited) sensitivity to sugar can determine the behaviour of forager bees: "sweets lovers" prefer collecting nectar rather than pollen. Beekeepers have noticed that the bees from one colony for some reason found the source of nectar quicker than bees from surrounding hives.

It is impossible to explain all aspects in the life of a bee through the genotypic differences of the subfamilies and their genetic variety. As to the phenomenon of age polyethism – the key inherited factor of social insects' success – there is no doubt that we should mention its flexibility. It was discovered that bees could change the polyethical group more quickly or more slowly and even "turn back", depending on the external or internal condition of the colony. Precocious species at the age of 7–10 days performed the function of forager bees in a colony consisting of young worker bees. The delay in changing the category of nurse bees to forager bees can be seen in swarming, because after that the colony lacks young bees (they will only appear in 3 weeks after the swarming) and the need for royal jelly is high. That is why old bees have to feed the larvae. In the case of the removal of all young bees, even forager bees whose royal jelly-secreting organ is non-functional, have to feed larvae. These glands revive again and, as a result, forager bees retrieve a second youth (the "physical rejuvenation" of old bees).

In the light of these facts (and, moreover, in the light of facts not mentioned here) an appropriate question appears: how do bees exchange information and coordination in the family? Order in the colony is impossible without these factors. Changes in the behaviour of the majority of the species (for example, when old bees begin to do non-typical work) and decision-making demand special signals or information.[3] Unfortunately, as is already known, nobody has yet tried to estimate the information flow in a specific colony and expressed it by the number of bits per time unit.[4] We can only imagine how difficult this is to do. The method of exchanging information between members of the colony is not so hopeless, but remains a question to be answered.

One of the best known ways of exchanging information in the family is the waggle dance performed by forager bees on the combs or even on the back of the sisters swarming in a specific place. Information is also exchanged by sounds: worker bees can make a sound with their wings in certain conditions. When a young queen leaves the queen's cage, she usually "sings," and other mature queens who fail to get out of the cell respond with her singing. A young queen or worker bees kill these potential queen rivals. That's why the sting of the young queen is ready to fight.

Alongside all these mechanical and acoustic signals, chemical substances are used to exchange information. It is natural that for a chemist this element of the colony's "information technology" is the most interesting. As to the waggle dance, it is described properly not only in Karl von Frisch's original works (ref., for example, von Frisch, 1965), but in other educational and popular books (Woyke, 1998; Krivtsov *et al.*, 2007; Tautz, 2008).

Honey bees have the most complicated and developed system of chemical communication provided by 15 glands (Free, 1987). These glands are developed in different ways and produce special substances depending on the caste (queen, worker bees and drones). Let's start with the queen.

---

3 *The problem of collective decision-making and information for making such decisions is reported in the special issue of the magazine Apidologie (vol. 35, 2004).*

4 *If somebody thinks that it is impossible, I would remind them that the Russian scientist Prof. V. Gorshkov managed to estimate the information flow for the entire biosphere of the Earth and for human civilization in 1035 and 1016 bits per second, respectively (the difference of 19 points is impressive, isn't it?!)*

2.2. Chemical "levers of power" of the queen

The queen is not adapted to gathering food and doing construction work because of her short proboscis, lack of tools for gathering pollen and wax glands; it cannot clean herself from its excrement and she would be doomed to death if the worker bees didn't help it. It is worker bees that feed the queen and take care of it – clean, remove its excrement, prepare the cells for the oviposition that will be inspected by the queen. Queen bees really are treated like royalty and there are about 10–12 worker bees near the queen. But the list of worker bees in the retinue of the queen changes in comparison with the queen while it moves in the combs. Except during the mating flights (as well as in swarming), the queen spends all its life in the hive surrounded by unceasing concern. But worker bees do not worship the queen all the time: they turn the young queen out of the nest in her first mating flight. They stop feeding the queen just before swarming, mercilessly pinch her and make her run so that the queen will lose weight and start flying.

If the queen is young and healthy, she is the only sexually productive female in the colony. If the queen disappears – dies a natural death from disease or for some other reason – the bees get worried. This is understandable: if they do not have a new queen the whole colony is doomed to progressive quick death.

Even from this short description of the relationships between the queen and the other members of the colony we can draw the simple conclusion that there are some communicative connections between them. In some way, the information about the presence of the queen in the hive – and even the information about her fertility – is spread to the other members of the colony (this fact has been discovered recently), and it makes the worker bees take care of the queen. This is a signal to the others that in the colony everything is in order and they can do their work.

So how is this information spread? Beekeepers and scientists studying the lives of bees and other collective insects noticed a long time ago that members of the queen's retinue touched the queen with their antenna and licked her with their tongue. They came to the conclusion that the members of the queen's retinue obtained some "signal" chemical substances, called pheromones.

Pheromones *(from Greek φέρω - "to bear" and ορμόνη - "impetus") is the collective name of chemicals secreted by animals in the environment, providing chemical communication between the same species. They are biological markers of the species, chemical signals that control the process of development as well as other processes connected with social behaviour and reproduction.*
Pheromones can be roughly divided into two main types: releaser pheromones and primer pheromones. The first type stimulates the individual to immediate actions and is used for attracting mates, giving a sign of danger and stimulate some other activities. The second type (primer pheromones) is used for forming definite behaviour and for influencing the physiology of species for their development.

At the present time, scientists have succeeded in finding the chemical structure of the pheromones of more than 1,000 species of insects and almost all of them belong to releasers: from the chemical point of view, primer pheromones have been characterized in only of one type of social insect – the honey bee.

The pheromones of the queen have several effects:

- they attract drones during mating;
- the attract worker bees in dividing the family by swarming;
- they suppress the sexual development of worker bees and as a result prevent them from emerging as laying workers in the colony (pseudo-queen);
- they identy the queen as the object of the worker bees' care.

The components, biosynthesis and functions of pheromones have been studied for about 50 years, but the work is not at an end. It was found that the pheromones in queens were produced by the mandibular and some other glands. With the help of different surgical devices, scientists extracted these glands carefully from the queen and identified the dedicated substances. First, they managed to determine the structure of one compound – *trans*-9-oxo-2-decenoic acid (9-ODA), which was called the "queen's retinue pheromone". It has become evident that the queen's signal complex consisted of five components: the previously discovered 9-ODA is completed by two enantiomers of *R*- and *S*-*trans*-9-hydroxy-(*E*)-2-decenoic acid (9-HDA), *p*-hydroxy methyl benzoate (HOB) and 4-hydroxy-3-methyoxyphenylethanol (HVA). This blend is called the queen mandibular pheromone (QMP).

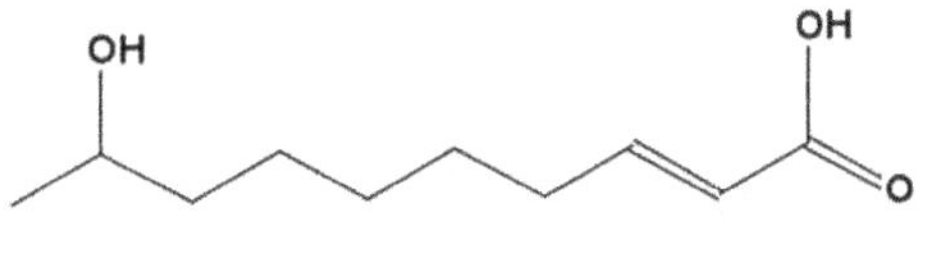

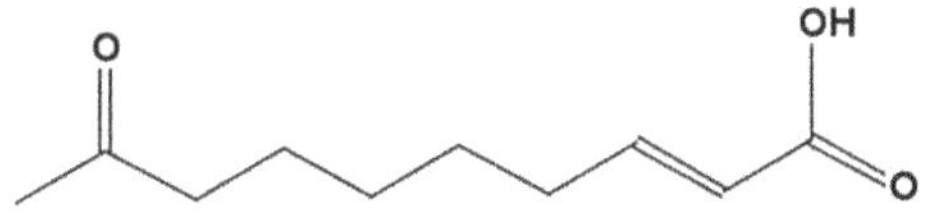

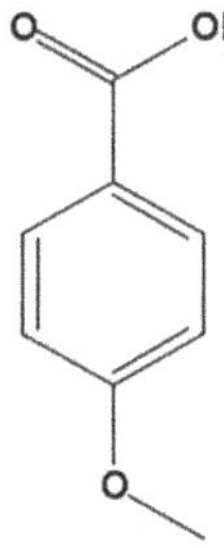

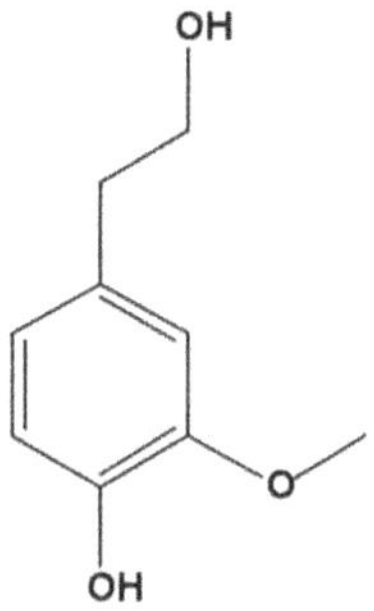

1 2 3 4

Figure 2.2. Chemical formulas of QMP components. 1 - 9-HDA, 2 - 9-ODA, 3 –HOB, 4 – HVA.

This pheromone is a typical primer, as it suppresses the maturation process in the reproductive system of other female bees and prevents the development of maternal instinct.

On the other hand, it serves as a releaser giving signals to the nearby working bees for them to realize that not far away there is an object to take care of. Researchers from Canada examined the so-called "attractiveness" to working bees, such as a whole mandibular gland extract of the queen as well as different combinations from the above-mentioned compounds (Slessor *et al.*, 1988).

Small amounts of these components were applied to pseudo-queen made of glass and the reaction of working bees was video-taped: at 30 s intervals the number of workers contacting the pseudo-queen was recorded. It appeared that the most attractive mixture for bees was a combination of the five QMP components: the activity accounted for approximately 69% of the activity involving the whole-gland extract, which was not split.

Second place was occupied with the mixture of 9-ODA + HOB + HVA (accounting for 53% of the activity around the queen extract). Approximately 36% of working bees expressed a desire to have contact with the "queen" on which an equal quantity of 9-ODA and HOB was applied. There was only a 25% response to the pure 9-ODA.

In another series of experiments, these researchers compared the number of "queen suites" on five beehive frames with a laying queen and the same quality of frames with pseudo-queens which were treated with whole-gland extract or a

9-ODA + 9-DA mixture. In the case of the whole-gland extract the response of worker honey bees to the pseudo-queens was 60%, and in the second case only 14% of the bees were attracted by the real laying queen. From the obtained result it was possible to draw the conclusion that the mandibular gland complex of the queen that serves to form the "retinue" (queen retinue pheromone, QRP) is more complex than it was supposed to be. Actually, four more compounds were found, which, in combination with the five previous QMP components, have the same attractiveness to working bees as the whole-gland extract (Keeling et. al., 2003). Three of these substances (methyl oleate, linolenic acid, and 1-hexadecanol) were derived from extracts taken from a queen's surface. Coniferyl alcohol is the fourth compound and it was isolated from queen mandibular glands. All four substances are low-volatile and are passed from a queen to a working bee by contact. The authors particularly noted the fact that these newly discovered retinue pheromones were not involved, either on their own account or in combination with QMP components, in the suppression of workers' ovary development and egg laying. This means that their role is to form a suite for a queen and, possibly, they may be involved in worker orientation during swarming. It is supposed that QMP has an influence on brain structure and thousands of genes in the brain; it activates genes that are associated with the work of the nurse bee and suppresses the genes associated with the work of the forager bee. Nevertheless, those genes under the control of pheromones have not yet been identified.

HVA plays a mysterious role because its content in the whole pheromone complex is rather small (approximately 200 times less than the content of the main component 9-ODA). It appears that this compound in particular suppresses aggressive behaviour of working bees towards a queen bee. Due to HVA, the level of one of the neuromediator dopamines in the brains of early bees is reduced (Beggs et al., 2007). Therefore, the real queen allows working bees to distinguish her from potential "usurpers" that impersonate the queen with the help of the queen pheromone imitation.

Rapid development of chemical analysis technology made it possible to separate compound mixtures of different classes and obtain new data about the content of queen mandibular gland secretion. In the late 1990s, German researchers identified more than 100 compounds in the mandibular gland extraction of the *Apis mellifera carnica* queen bee, and they also ascertained that their content changes as the queen ages (Engels et al., 1997).

According to chemical nature, one can classify the identified compounds into several groups. The most identified were linear alkanes with carbon atom numbers in a molecule from 7to 37 (25 compounds). They were attended by C_{12}–$C_{32\ \text{methyl alkanes}}$ (16 compounds) and linear C_{12}–$C_{39\ \text{alkenes}}$ (15 compounds). A significant group consists of the esters of saturated and unsaturated acids from hexylacetate to tetradecyl octadecanoate. Saturated and unsaturated carboxylic C_8–$C_{20\ \text{acids}}$ are represented by 10 compounds. However, the most interesting substances are bifunctional compounds: oxo- and hydroxy acids. Except for already known acids of this group (9-ODA и 9-HDA) and aromatic compounds (HOB и HVA) in the mandibular gland secretion, the following substances were identified: 4-hydroxybenzoicacidand 4-hydroxy-3-methoxybenzoic acid; 4-hydroxy-3-methoxyacetophenone; 3-hydroxyhexanoic acid,3-hydroxyoctanoic acid; 3-hydroxydecanoic acid; 7-hydroxyoctanoic acid; 9- and 10-hydroxydecanoic acids; *trans*-7-hydroxy-2-octenoic; 7-oxooctanoic; 9-oxodecenoic and 11-oxodecenoic; and *trans*-10-hydroxyl-2-decenoic acids. The last acid (10-HDA) is considered to have the most useful properties in bee products and we will discuss this matter in another section of the book.

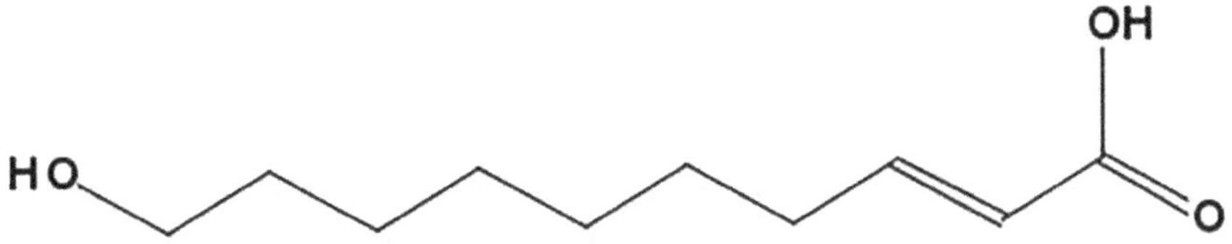

Figure 2.3. Chemical formula of trans-10-hydroxy-2-decenoic acid

These researchers also ascertained that the general content of extractive substances in glands increases while the queen develops, reaching its maximum by the 15th day of life, while the maximum content of 9-ODA is reached by the 10th day after the queen comes into being. At this age, the virgin queen bee makes her first mating flight! That is why one can consider that 9-ODA performs the role of attractant (sex pheromone-releasers) that attract honey bee drones. Indeed, special receptors are found in drones' antennae that are capable of perceiving only this unsaturated oxo-acid (Kaissling & Renner, 1968; Wanner et al., 2007).

***Receptors** (lat. recepere – accept) are special organs and cells that receive signals from the external and internal environment. The following types of receptors in organisms of animals can be specified as the main ones: chemoreceptor – reaction on chemical substance (responsible for taste and smell); photoreceptor – reaction on light; thermo-receptor – reaction on temperature; mechanoreceptor – reaction on different types of deformation, pressure and gravity.*

Receptors transform the energy of a stimulus into an electric signal that stimulates neurons. The mechanism of receptor stimulation is connected with cell membrane permeability for potassium and sodium ions. When the stimulation reaches its threshold, the sensory neuron is stimulated and sends impulses to the central nervous system. Insects have a low threshold of sensitivity. To stimulate the sensory neuron of the silkworm moth (Bombyx mori), for instance, it is enough the concentration of pheromone only has to reach 100 molecules in 1 cm^3 of air (compared with 1 cm^3 of air contains about 10^{18} molecules of nitrogen and oxygen). Such sensitivity is attained due to that fact that an insect has more receptors: on each of two antennae of the silkworm moth has at least 17,000 of them.

Evidently one more substance, oleic acid, plays a role in attracting honey bee drones during the mating flight. Its content in the mandibular gland extract of the queen reaches its maximum by the 10th day of living and after that reduces twofold. This acid probably plays the role of a substance that facilitates the distribution of ODA in the air and having, like the rest of the signal complex compound, low volatility.

At first, the content of 9-ODA and 9-HAD in the mandibular gland of the newly fecundated queen is very low but it increases rapidly. The chemical composition of glands becomes more complex: the emergence of methyl 4-hydroxybenzoate (HOB) and later 4-hydroxy-3-metoxyphenyletanol (HVA) has been registered at the age of 15 days. German researchers (Engels et al., 1997) believe that these aromatic substances are typical for fertile and active laying bee queens.

Components of queen mandibular pheromone (QMP) are low-volatile, which is why in hive conditions they can spread, in the form of gas, from the source of emission to a distance equal to approximately 12 mm (Kaminski *et al.*, 1990). In other words, this emission only has an influence on bees near a queen. Nevertheless, the QMP components circulate the entire hive and that happens thanks to the transmission from one working bee to another through physical contact among the insects. This includes the licking and antennating behavior, which signals the presence of a dominant reproductive queen and thereby establishes and stabilizes the "social fabric" of the colony.

Hence, the behaviour of the entire bee family depends on how active a queen is in producing QMP. A high level of QMP production suppresses the wish of bees to nurse a new queen, but as soon as the level becomes less than some threshold values (for example, when a queen becomes old or ill), working bees immediately start build up additional bee cells with young bee larvae, turning them into emergency queen cells. They act similarly when the bee family becomes too large and the hive is "overcrowded", which makes the circulation of QMP components difficult. In this case, the level of QMP per each bee lowers the necessary threshold value and the rearing of a new queen becomes a preparatory stage to swarming (Watmough *et al.*, 1998). On the edge of the cells the bees build a new type of queen cell – clusters.

The fact that the presence of a queen and level of her fertility is felt at the same time by the many thousands of bees in a family made some scientists think that the information within one hive can be transmitted not only by contact but also by air.. Evidently the first attempt (unfortunately not fully successful) to reveal the extract of volatile signal compounds made by a queen and working bees was undertaken by Moritz and Crewe (1991). These researchers managed to register more than 30 compounds in the volatile emission of queens, four of which were not present in the excretion of working bees. The scientists were unable to identify these unique components.

The task of identifying insect pheromone is rather difficult as these signal components are produced in small amounts. In order to obtain an extraction of 4 milligrams of bombykol (a pheromone released by the female silkworm moth), a group of German researchers headed by Adolf Butenandt had to dissect tens of thousands of female silkworm moths! But to evoke an appropriate reaction in insects only a very small amount of volatile pheromones are needed and this is due to the fact that insects have supersensitive chemoreceptors. It was determined that that just one pheromone molecule is enough to stimulate a sensitive receptor cell on an insect antenna. (a male can be attracted from 10 km distance).

Quite recently, American researchers repeated another attempt to identify volatile queen pheromones of the honey bee with the help of modern analytical techniques (Gilley et al., 2006). They compared the composition of volatile substances released by queens of different ages (virgin, newly fecundated and "mature" queens) and by working bees. As a result of all the analyses, the scientists registered four components that are released only by queens, not by working bees. One of them is terpene hydrocarbon ß-ocimene, released in larger amounts by "mature" queens, whereas the rest of the compounds are produced by virgin queens. Of all the other three components, one was identified as fatty aromatic alcohol 2-phenylethanol while the other two remain unknown. Unfortunately, the researchers' conclusion regarding the signal function of the identified compounds was not supported by the analysis of the working bees' reaction behavior.

We would like to draw attention to the fact that both identified "specific" compounds are, in reality, rather widespread components, which are released into the atmosphere by many species of plants (Isidorov *et al.*, 1985). They are also found in bee products, such as in different types of honey and propolis. That is why they should be constantly maintained in greater or smaller_quantities, not only in environment but in the air within a hive as well. Thus, the question about regarding the volatile signal components of a queen bee is a matter for future investigation and awaits enthusiastic researchers to find the answer.

2.3 Info-chemistry of worker bees

As a result of research held during recent decades it became possible to reveal the so so-called "chemical background" of workers' behavioral responsese. As is already evident, the age polyethism of worker bees is not absolute and is characterized by a known flexibility. Empirical observation and specially planned experiment results show that necessary information is transmitted with the help of primer pheromone.

For a long time, the above-mentioned delay in "aging" of hive bees in strong families, and their unwillingness to become foragers, even though their anatomical and physiological state would allow them to fulfill this kind of work at the age of 5–6 days, was inexplicable. Such a delay is observed in families where there are enough foragers. At the beginning of the 90s, American scientists experimented in an attempt to explain this mystery (Huang & Robinson, 1992). In the first part of the experiment, old and young bees from one family were separated from one another by a partition that gave them the opportunity for contact. The bees used this possibility actively to give (from the older bees to the young ones) nectar, and to lick and touch one another by antennae. In another part of the experiment these two groups were separated by a double partition that excluded any physical contact. It turned out that the young bees, when entirely isolated, began to mature more quickly (that is, in their behaviour).

Hence, it was concluded that a factor prevented from being passed on to the next age group was a primer pheromone that came down from the old bees to the young ones in two ways: through direct contact or in food. Only 12 years later this pheromone was determined as being ethyl oleate (EO) and it is transferred with food by way of trophallaxis (Leoncini et al., 2004).

Trophallaxis *(from Greek* trophe *– food and* allaxis *– exchange) is the transfer of food or other fluids among members of a community through mouth-to-mouth. It is most highly developed in social insects such as ants, termites, wasps and bees. The term was introduced by the entomologist William Morton Wheeler in 1918.*

The fact that OE plays the role of such pheromone is proved by its content, which was two and a half times higher in full extracts from forager bodies than from bee nurse bodies. Moreover, in the forager foregut (which serves for the storage of nectar and honey) the quantity was 30 times more EO than did "nurses". Small quantities of EO were also detected on the cuticles of foragers, whereas nurses had undetectable levels. In researches with glucose labeled with carbon atom ^{13}C, it was determined that the foragers biosynthesize EO *de novo* in the honey crop.

The conclusion is evident: old bees delay young bees moving on to the category of foragers (at least partially) through the transfer of EO by way of trophallaxis. Reducing the EO circulation level in a family as a result of forager levels reducing, leads to the acceleration of the "aging" process of young bees and this is an adaptive reaction to changes in the inner conditions within the family. Perhaps EO functions together with unidentified components that are produced by foragers: multi-component pheromones are common among social insects. One more important conclusion is the following: trophallaxis is one of the main information transfer mechanisms (communicative channel) that is carried by primer pheromones.

The number of worker releaser pheromones includes an alarm pheromone (that signals threats) and a sting pheromone, which stimulate an attack on an enemy. The behaviour of the alarmed bees is different from their behaviour during defense attack. The alarmed bees begin to pose aggressively and rush toward the source of the potential danger, the locality of which is marked with the volatile pheromone of anxiety. When a bee stings an enemy, components of the second pheromone begin to be produced, which induces other insects arriving to attack.

The first of these releasers is produced by the mandibular glands and the second by the venom glands (the sting sheath glands) of the workers, as well as the Koschevnikov gland. On the whole, about 40 compounds are identified in workers' stinging organ extracts, however a major part of these are precursors and transition products of the biosynthesis of the main components that are carried out in the bee venom gland (Blum & Fales, 1988). This releaser also consists of a mixture of low-molecular alcohols and their acetates. In 1974 Blum and his co-authors determined that mandibular glands emit ketone 2-heptanone (methyl amyl ketone) and the main components of the second pheromone are the ester isopentyl acetate (IPA) and unsaturated alcohol *cis*-11 eicosen-1-ol. In addition, it has been supposed that this odorant substance attracts the European beewolf (*Philanthus triangulum*). This wasp, which lives in an earth hole, can inflict serious damage on bee families (Schmitt et. al., 2007).

Isopentyl acetate is a volatile substance and is easy to spread in the air, which is why it is sometimes referred to as a target-marking pheromone. Evidently *cis*-11-eicosenol prolongs the activity of the more volatile IPA, presumably by slowing down its evaporation (the blend of IPA and eicosenol is active for a longer time than IPA alone). In other words, this alcohol is a smell fixative like ambergris, which is widely used in perfume production.

Other low-volatile compounds can be used as a fixative. For instance, in a poisonous substance of an Indian giant honey bee (A. dorsata*)* cis*-11-eicosenol is not found but 2-decen-1-ol acetate is. Bees of this species are the most aggressive ones; they are ready to rush at any point from their nest to attack an approaching enemy. It is probably these bees that were described by Rudyard Kipling in his book about Mowgli.*

It is interesting that Eastern honey bees (Apis cerana*), which have much in common with our honey bee, differ in character. They are less aggressive and are not inclined to pursue an enemy. It turns out that the stinging organs of this species of bees produce two times less IPA than* A. mellifera *L. and only small quantities of alcohol* cis*-11-eicosenol, although the content of the last compound in a poison sac of* A. cerana *is rather large (Schmidt et al., 1997). But this does not mean that the Eastern bee is less protected, it is simply that it has worked out another more effective defense strategy. For instance, the bees when alarmed in their hive make a specific vocalization that resembles the sound of a hissing snake ready to attack, and such a sound frightens many honey-loving animals.*

It was also reported that aging mandibular glands produce 2-heptanone in larger quantities and the maximum quantity is reached in foragers. This substance has the quality of a weak repellent and probably serves them as a repellent forage-marking scent to mark visited flowers (Vallet *et al.*, 1991; Giurfa, 1993). In this way a bee warns its sisters: – *I have been here and have taken everything, fly on!* Indeed, if for a few minutes one watches bees gathering nectar, for example from s linden tree full of flowers or any other tree in bloom, they will confirm this fact. They will see that a bee aiming for a specific flower suddenly changes its direction of flight and rushes to another flower that seems to be no different from the previous one on a nearby branch.[5] But how much time and energy is saved by thousands of foragers due to such a "message"! It has also been suggested that 2-heptanone is produced by bee guards and to frighten off robber bees.

A contrary role is performed with a pheromone that is produced by the Nasonov

---

[5] *It is an interesting fact that bees avoid visiting flowers that have been visited by bumblebees and vice versa (Stout & Goulson, 2001). At the same time bees make mistakes in such behaviour more rarely than bumblebees. Evidently this can be explained by the relative volatility of the repellent forage-marking scent of these two insect species: 2-heptanone has a low boiling temperature, therefore it is easy to volatilize and faster to disperse, whereas the pheromones of bumblebees consist of long-chain alkanes and alkenes which have high boiling temperature.*

gland of worker-bees: it serves to mark the places which are rich in nectar plants. Some authors believe that this pheromone is especially important to mark those objects with a weak scent, such as a water source and plants, where nectar does not have strong scent (Schmidt, 1999). Secretion of the Nasonov gland consists of seven biochemically related terpene compounds: two isomers of terpene aldehyde citral, two terpene alcohols (geraniol and nerol), two isomers of terpene acids (geranic and nerolic acids) and sesquiterpene alcohol (E,E), farnesol. This pheromone serves in orientation: it helps the bees to find their nest whether it is a hive, or the hollow of a tree trunk. Some of the bees returning from flight very clearly pause before a nest entrance: they raise the tip of their belly and stretch it out, simultaneously fanning the flared gland through rapid wing movements. As a result, the secretion volatilizes from the revealed groove of the abdominal tergite (in normal conditions this groove is hidden by the sixth and seventh abdominal tergites). Such excretions also help lost bees to find their family during swarming and excite them to form "swarm clusters".

As we can see, a rather branching system of information operates in the bee family, with the help of chemical substances, which are produced by many glands. In addition, the composition of the produced compounds is rather specific to the different castes. Workers' mandibular glands produce 2-heptanone and royal jelly, whereas the same glands of the queen bee produce a special complex of components that does not allow the workers to participate in making brood (QMP), compelling them to serve their queen faithfully (QRP). Another example of a caste-specific character are the Dufour glands: if workers' secretion consists of five odd-numbered $C_{23}H_{48}$–$C_{31}H_{64}$ *n*-alkanes, in this case in the secretion of a queen, in addition to the above-mentioned compounds there are also 28 substances that include large quantities of esters of fatty acids and alcohols with a chain of 14 and 16 carbon atoms (Katzav-Gozansky et al., 2002). Many researchers find the role of such secretions mysterious. Some of them are presumed to serve as a pheromone marker for egg laying by a queen, but this theory is not yet scientifically confirmed.

Alas, as in other complex systems, within the honey bees' system of chemical communication different failures may occur. No matter how effective the mechanism of suppression of the worker reproduction instinct, there are, in all "ordinary" families, such workers that regularly lay eggs (Ratnieks & Visscher, 1989). The majority of such unfertilized eggs are killed by the so-called "police", although sometimes these egg-laying workers manage to deceive the "police" (Oldroyd et al., 1994; Martin et al., 2002). It turns out that they not only lay eggs but also produce in their mandibular glands a mixture of substances that are similar to queen pheromone (Crew & Velthuis, 1980).

The Dufour gland secretions of such "queen-pretenders" are also changed: in some of them, the above-mentioned complex ethers a present (Katzav-Gozansky et al., 2002; Dor et al., 2005). Moreover, these pretenders acquire their own "suite" and are capable of suppressing the development of the ovaries of the bees that surround them! Very often, and not in just one example, such pseudo-queens have appeared in the families of the Southern African race of honey bees, *A. mellifera capensis*. It is significant that egg-laying workers in these families manage to coexist with one another and do not challenge one another to duels which end in the death of one of the participants (Moritz et al., 2003a). However, real queens in the families of this type race are nervous about the presence of such "pretenders" and they try to get rid of them. As a rule, a true queen manages to do this, but sometimes the duel ends with the death of a queen (Moritz et al., 2003b).

Not long ago, Israeli scientists (Malka et al., 2007) reported one more interesting observation. It seems that the process of becoming an egg-laying worker in the European bee race is a reversible process. After being placed in a "normal" family with a strong queen, the functions of the ovaries and glands that used to produce unusual pheromones of such workers are eventually suppressed.

Moreover, it has been noticed that sometimes laying workers show responsibility for the whole family and want to redeem their fault by building a queen cell. Alas, unfertilized eggs, even in queen cells, and even having been generously fed with royal jelly, eventually only develop drones. Although such families can be saved if on the frame in the middle of the nest to inoculate several queen bee cells taken from normal strong families that are ready, for example, to swarm. Bees usually accept such queen bee cells and derived from their own nest queens. Egg-laying workers will continue to lay eggs until the moment when in a family appear brood of "foster" queen.

2.4. Pheromones of larvae cry out: We are here, feed us!

One more group of signal substances informs all members of a bee family about the brood, its status (whether a worker or a queen or a drone develop from a larva), stage of larvae development and necessary regime of feeding. That is why it is called a brood pheromone. In the words of Doctor John Borden (Borden, 2011), with the help of these substances larvae give a signal to the workers: "we are here and we are hungry!" In the content of the signal complex that was extracted from the surface of larvae, 10 compounds were identified and all of them turned out to be methyl and ethyl esters of higher C_{16} and C_{18} fatty acids such as: palmitic, stearic, oleic, linolenic and linoleic acid (LeConte et al., 1990). These esters, produced by the salivary glands of larvae (LeConte et al., 2006), are not volatile, some of them are derived from unsaturated acids and are non-stable as they are easy oxidize.

During special experiments the authors determined that these esters in different combinations fulfill different roles. For example, a mixture of methyl linolenate and ethyl palmitate has an influence on the endocrine system of workers, and in combination with methyl linoleate and methyl oleate serves as a signal to seal the cells of six-day larvae that are ready to become pupae. Methyl palmitate and ethyl oleate stimulate the development of workers' mandibular glands and protein production, while ethyl palmitate and methyl linolenate suppress the development of workers' ovaries (Mohammedi et al., 1996; 1998). Thus, aliphatic esters of the brood have an impact on workers' behaviour and their physiology, which appears to be as both a releaser and primer pheromone at the same time.

Beekeepers know quite well that the peak of brood growth occurs at the same time as the peak of bees gathering pollen, which is so necessary for the production of royal jelly, which serves as food for the young larvae. It is evident that the more larvae there are in the open cells, the larger is the total quantity of brood pheromones produced. In special experiments with a synthetic mixture of esters that makes brood pheromone, it has been shown that it is precisely this pheromone that makes bees gather pollen (Pankiw & Page, 2001).

2.5 The drones "Men's club"

Again the author can only guess why scientists pay less attention to the male population of a hive. Nevertheless, the mode of life of these "flying gametes" makes me think that chemistry signalizing is extremely important to the fulfillment of a task set by nature: to sustain generation growth and maintain the species *A. mellifera.*

For successful mating a drone needs to find a queen and the signal substance 9-ODA, which is produced by a queen helps them to do this. However, drones do not fly around chaotically trying to catch molecules of this attractant in the air with their antennae, with the help of which they manage to fly in the right direction to find the queen.

Beekeepers noticed long ago that mating does not happen everywhere. Drones constantly gather at a definite height (20–30 metres above the earth) and at a definite place that is called the "drone congregation area". This is a "spot" whose diameter measures from 50 to 200 metres. It has also been established that drones can gather at the same place year after year and over a number of years (Koeniger & Koeniger, 2000). They gather at this place (sometimes drones can cover a distance of 5 km) even if there is no hope of a rendezvous with a queen. It is nothing less than a "Men's club"! In this "club", policy is rather democratic: it is open to members from different families. As German authors reported (Baudry et al., 1998), in one such "drone congregation area" in Germany, drones from 238 colonies were registered at the same time. Membership in one and the same "club" is not constant: if there are several congregation areas within a 5 km distance, they can be attended in turn. Although democratic behaviour mostly concerns "equal drones": if there are bees of different breeds on such a territory, then the drones of each of these breeds, although collected in the same cluster zone, are kept separately at different heights. For example, drones of *Apis mellifera carnica* prefer to be in the upper levels.

How many drones are at the congregation area at the same time? The answer to this question was found not long ago. Five years of research showed that approximately at one area 11750±2145 individuals gather (Koeniger et al., 2005). Does this mean that this horde rushes to a queen? Of course not. According to the data of these authors, only 20–40 drones enter the "breeding comet" that involves competing with one another for a queen. The winners are those drones to whom nature has given the strongest wings, the best sight and the highest sensitivity to 9-ODA.

Let me draw to a conclusion and try to put forward some hypotheses concerning the signalling substances of a honey drone bee. Firstly, if drones are gathered in large quantities at a specific place and in a limited air volume, it is logical to admit that there are some volatile signal substances that let them find one another, in other words an aggregative pheromone. Secondly, the fact that bee drones of different breeds stand apart (at different heights), makes us think about the subtle difference in the content of these pheromones. Thirdly, the drone congregation area is not necessarily situated near the apiary or the habitation of wild bees. Hence one way or the other, a young virgin queen should find it and in this case a volatile attractant substance produced by some glands of the drones will be very useful.. However, the possibility that the queens are attracted by the same substances that serve drones as an aggregation pheromone cannot be excluded.

The Israeli scientists (Lensky et al., 1985) have proved the existence of an aggregative pheromone that is produced by drone mandibular glands. They have determined that the extract of these glands, put on a "drone dummy" (which was raised to a height of 8–12 metres with the help of an air balloon), attracted drones flying in the air. However, the chemical content of this pheromone has not yet been discovered.

The overview of signaling substances of the honey bee given in this chapter is far from complete. This is largely because of the fact that emissions of not all 15 bee glands are investigated. That is why I prefer to follow the principle of Ludwig Wittgenstein: "what is not clear till the end should remain silent".

2.6. Pheromones at the service of beekeepers

The scientists that made a significant contribution to the study of honey bee pheromones were at the same time pioneers in the practical use of the knowledge they had obtained. Many of them patented the composition of synthetic pheromones and some part of these patents was implemented into practice. For example, the American authors who decoded the composition of queen pheromone offered a synthetic mixture to the "formation and stabilization of the social factory" of honey bees (Slessor et al., 1991). Not long ago in the USA and some other countries in Western Europe, there appeared a synthetic product for sale under the name of "Bee Boost" (produced by the company "Phero Tech Inc.", Canada) which contains a blend of five main components of queen pheromones which we have spoken about in the chapter 2.2. It is used to "calm" a family that for one reason or another has either lost a queen or it is necessary to change her. The use of this product prevents bee escape, lays an emergency queen cell and detains swarming for a long period. This product is especially recommended for bee transportation without a queen.

The same company produces another synthetic product under the name "Swarm Catch", which serves to attract and catch a swarm. Some American beekeepers call it the "come-hither pheromone". A prerequisite for its production has been the decoding of the composition of the secretion that is produced by the odorous Nasonov gland.

It is a well-known known that swarming is one of the two main factors that has an impact on bee family breeding (the other factor is wintering). Although the problem of swarming that is connected with the shortage of honey at the present moment concerns small non-professional beekeepers/apiaries where planned bee breeding is not practised by way of an offshoot. The owner of a small apiary is not always able to keep his eye on when a family is preparing to swarm, and not always capable of finding a flown cluster, particularly if he/she lives far away from it. That is why so many products for catching swarms are produced. Besides those products already mentioned, there are other synthetic pheromones produced by one Serbian company "Eurotom" (its products are registered in many other countries). The Russian company Agrobioprom for these purposes produces "Apiroy", "Uniroy" and "Apimil" (the last one is recommended when it is necessary to put in a new queen).

A positive trait/feature of all these mixtures is that they contain only natural components. But what about their effectiveness while catching swarms? I have data about only one American product "Swarm Catch". With this product one can bait 50–80% of flown clusters. This is not so few if we take into consideration that the effect of controlling a "catch" (without the use of such bait) amounts to only 21%. But I remember that my grandfather managed to catch 100% of cases. I remember this because when I was a teenager I used to help him take beehives off a spruce and drag them 600–800 metres to a house. The secret of success is very simple and indeed it is not a secret for many beekeepers: before hanging a hive on a tree, my grandfather diligently rubbed the inside the hive with melissa oil. The leaves of this plant (*Melissa officinalis*), which is also called "lemon balm", contains all the main volatile components of the Nasonov gland pheromones, namely citral, geraniol, nerol and their acids.

French researchers who examined the brood pheromone (LeConte et al., 1990) later patented the way of bee behaviour formation (essentially in queenless families) by way of processing larvae with different combinations of esters of palmitic and stearic acids (LeConte et al., 1997). One of the purposes of this development was to find the opportunity to stimulate the royal jelly production of nurse bees and to breed new queens. In their turn, American researchers who found the connection between the intensity of brood pheromone release, the quantity of royal jelly produced by nurse bees and the accessibility of protein food (Pankiw & Page, 2001), which for bees is pollen, also patented the composition of a synthetic pheromone (on the basis of the same esters) to increase its provision by foragers (Page & Pankiw, 2003).

I have never read about any differences in the content of brood pheromones that are produced by workers and drone larvae, although one may think that such differences do exist. The selective character of the invasive *Varroa destructor* into worker and drone larvae cells makes one consider such subtle differences. As has already been said, the latter are more often invaded by mites (approximately 15 times more invasions). It is difficult not to think that when choosing a cell, a mite is guided first and foremost by the scent of some specific volatile substances that are produced by drone larvae. If such substances do exist and if they can be discovered, they may be used in the biological struggle against these dangerous parasites.

Chapter 3 The mysterious royal jelly

3.1. Myths and secrets of royal jelly

Of all the products produced by worker bees, royal jelly (RJ) is the most admirable due to its unique properties. At the same time, it is also the most mysterious and therefore "overgrown" with various myths that wander from one literary source to another. In 1792, the Swiss explorer F. Huber named it "gelée royale". From French, the name migrated to English and became RJ. Presently, there are many completely different interpretations of its origin. Some authors associate its name with the fact that this rare and expensive product was supposedly available only to crowned heads, which is a typical myth. Others believe that it originated from the fact that it serves only as food for the queen bee. Furthermore, some believe that this is the only food consumed throughout her life, but this is also a myth.

It is now well known that the bees feed RJ not only to the queen but also all the larvae. Those of them who in due time will turn into workers and drones, receive it from the nurses only during the first three days of postembryonic development (after their hatching from the egg) and the larvae of future queens are fed this twice as long, until the moment of sealing the queen cell with wax lid. In *quantum satis* mode, larva of the future queen bee literally floats in RJ. Moreover, before the queen cell is sealed, the workers bring in such a large supply of RJ that it remains not fully used by the end of the completion of the pupa's development. During wintering, there is no need to feed the queen with RJ.

There is hardly any product known to man that has the same unique properties. This uniqueness consists primarily in the super-high nutritional value of RJ. Notably, 6–7 days after hatching, the weight of the larva increases 1500–3500 times. Neither cow's milk nor bird's milk from pigeons, flamingos, or male emperor penguins can be compared with RJ in this respect. For the first five days of life, a young virgin queen is fed with honey alone as it is the "energy" food necessary for mating flights. After mating begins, feeding is enhanced with RJ, which leads to the rapid completion of the development of her ovaries. Only thanks to RJ is the queen able to lay 1.5–2.5 thousand eggs per day, the total weight of which can significantly exceed the queen's weight. Such enhanced nutrition is necessary during the laying period but it is completely unnecessary at the end of it; therefore, in late autumn and winter, the queen, like all other family members, feeds on honey alone.

Another feature of RJ is its resistance to microorganisms. Its ability to inhibit microbial growth was first established in the 1950s by French scientists Remy Chauvin and Pierre Lavier, who noticed that it only became brown after several weeks of storage at room temperature.

Another unique property of RJ is its effect on the bee caste, depending on the feeding regime of the larvae. The larvae of future workers and drones are fed with RJ only for the first three days and rather sparingly as their daily ration is no more than 10–15 mg. After that, they receive only honey and pollen gruel from the nurse bees. Larva of the future queen are fed with RJ alone and in large quantities and its content in the queen cell reaches 250–300 mg.

3.2. Fateful food: The diet of the larva determines the future fate of the adult bee

Since the time of Adam Schirach, researchers agree that the diet of larva determines the caste of females, workers, and queens and thus their fate. However, the problem of the "diet–caste" relationship is far from being resolved and opens up a wide field for imagination. In some literary sources, for example, you can read that "jelly" for feeding worker bees is *somewhat different* in its qualitative chemical composition from "jelly" intended for the larvae of future queens. Some sources even say that there are two *completely different* types of "jelly", one of which is used only for feeding the larvae of worker bees and drones, called "bee jelly". It is thought to be secreted by the pharyngeal glands of nursing workers. The other type of jelly is RJ itself, which is used to feed the larvae of the queens and the queens themselves[6]. It consists of a mixture of secretions of the pharyngeal and mandibular glands of workers. At the same time, some nursing bees supposedly feed only the queen larvae, while others, only the larvae of worker bees and drones. As confirmation of this, data on the differences in the quantitative composition of bee jelly and RJ are cited. It is purported that in RJ there is much more juvenile hormone, pantothenic acid, biopterin, and neopterin, as well as a different ratio of glucose and fructose. Additionally, it has been reported to contain a unique compound, 10-hydroxy-*trans*-

[6] *There are publications that even talk about three types of jelly: for feeding queens (royal jelly, RJ), workers (worker jelly, WJ), and drones (drone jelly, DJ) (Schmitzová et al., 1998).*

2-decenoic acid (Fig. 2.3), which I already mentioned in the previous section.

In some scientific publications (Beetsma, 1979), it is argued that even the larvae of the queen bee during their development receive different food; for the first three days they feed only on the secretions of the mandibular glands and later on a mixture of secretions from the same and the pharyngeal glands in equal proportions. Future workers are fed with a mixture of the secretions of the mandibular, pharyngeal glands, and pollen in a ratio of 2:9:3 (Jung-Hoffman, 1967; cited from Winston, 1991). It raises the question of where these relationships came from. The answer to this difficult question is as follows: It is in such proportions that the food of the larvae of future workers contains a certain “white”, “transparent” and “yellow” substance. The first of them is the discharge of the nurse’s mandibular glands, the second is the discharge of her pharyngeal glands, and the third is... mainly pollen. It remains only to guess how the author managed to see these multi-colour fractions in an almost microscopic drop of liquid food put by the worker in the cell with the larva. After all, they are not served in separate packages to the larvae. Reports on these fractions and their ratios (albeit without mentioning their colour) have been appearing in publications from the late 1960s to the present (Beetsma, 1979; Winston, 1991; Peters et al. , 2010).

Adult queens from nursing bees “presumably receive mainly larval food, possibly with some addition of honey,” as reported in a fundamental work on bee biology (Winston, 1991, p. 62). It is unclear what exactly this “larval food” consists of. It may be from the “white” substance mentioned in the same book or from its mixture with a “transparent” substance. The “yellow” substance (pollen) is not mentioned in this source.

The queen is fed by young non-flying bees that are part of her retinue (Fig. 3.1), the composition of which is constantly changing. When the queen, laying eggs, moves from one cell to another (and in a day she can walk up to 250 m), young bees turn to her, forming a circle, touch their antennae, lick her and every 10–15 minutes are fed with RJ and at the same time behave quite persistently, literally imposing their services. According to Istomina-Tsvetkova (1953), during the breaks in laying eggs, she can receive food sequentially from several (up to five) young workers. The bees of the suite do not accompany the queen for long and after a while a new suite appears around her. The feeding is then continued by other bees who were previously engaged in feeding the larvae.

Figure 3.1. Queen bee, surrounded by her retinue (photo by S. Bakier, with permission)

It is difficult to imagine a radical change in the composition of jelly in a bee that has just fed brood larvae as soon as the queen is next to it. The opposite change would also have to occur, when the same bee leaves the entourage and returns to its previous occupation of feeding the larvae. It may be queried that *Apis mellifera* anatomists have somehow overlooked a kind of valve that opens and closes the mandibular duct in a nursing worker bee in response to the approach or removal of the queen. The constant change in the composition of the retinue, which is an established fact, also does not fit in with the idea that there are special nurses for the queen and special nurses for the larvae in the family, moreover, only future workers and only future drones. Then it would be necessary to assume the existence of a third category of wet nurses who are exclusively engaged in the care of queen cells.

There are also reported differences in the qualitative and quantitative composition of the above-mentioned compounds in bee and RJ. The mentioned difference in the content of juvenile hormone between the two types of "jelly" is hardly relevant as this hormone is produced in the body of the larva itself in a section called the *corpora allata*. It enters the hemolymph of the larva and plays an important role in its metamorphosis. In the larvae, from which the queen is formed, the activity of *corpora allata* is extremely high and therefore, the content of this hormone in the hemolymph is high (Cymborowski, 1998).

Further, analysis has not confirmed that 10-HDA is contained only in the larval food of future queens. According to our data, this acid is the main (50–

65%) component of the lipid part of the feed in jelly from cells with the larvae of workers and drones, as well as from the queen cells. We also did not find significant differences in the content or ratio of glucose and fructose in jelly from queen cells or cells with larvae of workers and drones (note: we took samples from the same frames with swarming queen cells and bee brood).

As for the quantitative differences in the content of pantothenic acid, biopterin, and neopterin, I have not been able to find scientific publications in which they would have been confirmed by chemical analysis with proper statistical processing of the results confirming the significance of these differences. In this case, natural (seasonal, regional, and other) fluctuations in the composition of RJ should be considered. Without such experiments, the division of jelly into WJ and RJ, and further into DJ seems to be unreasonable. Natural fluctuations in the composition of RJ can be very significant; this has been observed by myself in my own experience and reports by other researchers (Karaali et al., 1988; Zeng et al., 2011). Below are data on the composition of fresh and lyophilized (vacuum dried at a low temperature) RJ, which show the wide range of concentrations that individual compounds and groups of compounds can be present (Sabatini et al., 2009).

Table 3.1. Chemical composition of fresh and lyophilized RJ

Component	*Relative composition (%) of royal jelly*	
	Fresh	*Freeze-Dried*
Water	*60–70*	*<5*
Lipids	*3–8*	*8–19*
10-HDA	*>1.4*	*>3.5*
Protein	*9–18*	*27–41*
Sum of sugars	*7–18*	*22.7–30.9**
Fructose	*3–13*	-
Glucose	*4–8*	-
Sucrose	*0.5–2.0*	-
Ash	*0.8–3.0*	*2–5*

** Data from (Karaali et al., 1988).*

Some experimental data also contradict ideas about the different composition of jelly used by bees to breed different castes (queens, workers, and drones). It has been reported, for example, that only drones, not queens, were hatched from diploidal (i.e., produced from fertilized eggs) drone larvae, artificially fed with RJ from queen cells (Polaczek, 2001). Comparison of the protein composition in aqueous extracts of feed from cells with young (less than three days old) larvae of queens, workers, and drones did not reveal any differences (Patel et al., 1960). The same proteins in the same quantities were found in the food of older larvae of the queen with an age of 4–6 days. However, in food of the old larvae of workers and drones, the authors missed some peptides. This is not surprising, since upon reaching the age of four days, these larvae were transferred to another, honey-pollen diet, which did not include RJ. More recent studies, carried out at a higher experimental level, confirmed the complete identity of the main peptides of bee jelly and RJ (Schmitzová et al., 1998).

It seems to me that the level of our knowledge at the present time allows us to conclude that the caste of the future bee depends on two factors. First, it depends on whether the larva develops from a fertilized or unfertilized egg. This determines the sex of the future individual. Second, how long during their development did the larvae and prepupae of females receive RJ. With exactly the same qualitative composition of RJ, the amount of substances responsible for the development of the reproductive organs in females can be very different. For the simple reason that future workers are fed RJ for only three days (and even then, only sparingly) and queens their fill for six days, a simple calculation shows that the larvae of workers and drones eat no more than 35 mg, while the queen larvae and prepupae eat no less than 1200 mg of RJ. Therefore, the ovaries of the workers who were born are underdeveloped and the young queen is almost ready for mating flight and fertilization. It is another matter that we still do not reliably know which components of RJ make larva hatched from a fertilized egg into a queen and this issue is awaiting resolution[7].

Regarding the differences in the qualitative composition of RJ, it is now possible to say with confidence that it is species-specific; the development of *A. mellifera* larvae stopped when they tried to feed them with RJ of other species of bees, for example, the eastern bees *A. cerana* (Schönleben et al., 2007). However, this effect was not observed when the larvae are fed with RJ taken from bees not of their own race but belonging to the same species. For example, larvae of the South African bee

---

7 *According to some researchers, specific peptides in royal jelly are responsible for this differentiation (Kucharski et al., 2008; Peixoto et al., 2009).*

A. mellifera capensis developed well if they were given RJ produced by the workers of the North African bee, *A. mellifera scutellata*.

3.3. Royal jelly - a panacea for all diseases?

Nothing is necessary for everything the human race, as medicine.
Quintilian

For a relatively long time, chemists have discovered the main components of bees' RJ, which includes sugars, fats, proteins, vitamins, and mineral components. However, this information did not in the least satisfy physiologists and physicians who observed its effect on the human body. According to many of them, for these components to cause physiological and metabolic reactions in the body, their content would have to be at least ten times higher (Steyn, 1973).

The biological activity of RJ was noticed even before the study of its chemical composition began (McCleskey & Melampy, 1938). In the first half of the twentieth century, it was possible to divide RJ into several fractions, in which simple sugars (glucose and fructose), sterols, vitamin E, proteins, fats, and mineral components were found. In 1940, Canadian scientists Townsed and Lucas discovered a previously unknown $C_{10}H_{18}O_3$ organic acid in the ether extract of RJ and suggested that it was this acid that was the main physiologically active component of this product. In 1957, Butenandt and Rembold established its structure, which turned out to be 10-HDA[8]. This acid is called unique since it has not yet been found in other animal organisms or in any other natural products. However, this is not the only cause for its uniqueness.

The imagination of researchers has always been amazed by the resistance of RJ against various kinds of microorganisms. If we consider the high content of sugars, fats, and proteins in it, it becomes quite clear that only substances with a powerful antibiotic effect can protect RJ from the ubiquitous bacteria and microscopic fungi, for which it would seem to be an excellent substrate. There is probably no need to convince anyone that humanity is in dire need of such means.

---

8 *Published in German, Butenandt and Rembold's report went unnoticed by American scientists, who independently rediscovered 10-HDA in royal jelly two years later (Barker et al., 1959).*

The more than 80-year war against microbial infectious diseases, with antibiotics as the main weapon, has been rife with victories, albeit temporary ones. The enemy was able not only to survive but also to achieve revenge. The massive use of antibiotics has led to the emergence and accumulation of resistant forms of microbes. Dangerous human pathogens, such as Salmonella typhi (the causative agent of typhoid fever), Mycobacterium tuberculosis (the causative agent of tuberculosis), Vibrio cholerae (the causative agent of cholera), Pseudomonas aeruginosa, Enterococcus faecalis, and Staphylococcus aureus, among others, have already developed resistance to many of the antibiotics used against them.

Therefore, it is not surprising that the individual fractions and compounds isolated from microbial resistant RJ were investigated for their bactericidal properties. It turned out that it is 10-HDA that is active against many types of bacteria and fungi, almost not inferior in this respect to such antibiotics as penicillin and chlortetracycline (Blum et al., 1959). Interest in 10-HDA (and RJ in general) increased even more after Canadian scientists in a series of articles first reported the anti-cancer properties of this acid (Townsed et al., 1959; 1960; 1961). In *in vitro* experiments, the anticancer activity of RJ was later confirmed by other researchers (Kimura et al., 2003; Izuta et al., 2007; Nakaya et al., 2007).

Further biomedical research has shown that RJ has a surprisingly broad spectrum of antimicrobial activity. Among the microorganisms suppressed by RJ were the causative agents of such dangerous diseases as tuberculosis, anthrax, and salmonellosis, among others. Moreover, in experiments with the culture of *Staphylococcus aureus*, which can cause a whole series of serious diseases in humans (pneumonia, meningitis, osteomyelitis, etc.), the ability of RJ to enhance the bactericidal effect of honey was shown (Boukraa et al., 2008). When the same amounts of RJ were added to four types of monofloral honeys, their activity against this microbe increased by about 20 times. Interestingly, these mixtures were about twice as active as pure RJ.

Not only microbes, but also pathogenic fungi that are resistant to many antibiotics have been found to be sensitive to the action of RJ. For example, fungi of the genus *Candida*, which are often fatal to people with secondary immunodeficiency (AIDS patients, cancer patients undergoing chemotherapy, etc.), are suppressed by RJ in about the same way as the synthetic fungicide itroconazole (Melliou & Chinou, 2005). RJ, in this respect, is noticeably inferior to another synthetic fungicide, 5-fluorocytosine, but recently, *Candida* fungi have shown the development of resistance to this drug.

Antimicrobial peptides (AMP) are natural antibiotics with molecules of only 12–50 amino acids. They are synthesized in all living organisms in response to bacterial infection. The action of AMPs is based on their ability to attach to the surface of pathogen cells, disrupt the integrity of cell membranes and thus lead to their death. At present, more than 1500 AMPs have been described and high hopes are pinned on them. However, it is far from practical application of this discovery since they have not yet learned how to obtain them in sufficient quantities and with acceptable financial costs. But even with a successful solution to this problem, there is no hope that AMF will forever save us from infectious diseases. Their massive use would inevitably lead, as was the case of synthetic antibiotics, to the emergence of resistant forms of bacteria. Even before the use of AMP in clinical practice, it became clear that microbes can develop their own ways of defending against them. The main mechanism of this defence is the modification of the outer layer of the microbe's cell membrane, which excludes the possibility of AMP attaching to it. Many dangerous pathogens have chosen this strategy of protection (the "camouflage coat" strategy), including Staphylococcus aureus, Klebsiella pneumoniae, and Proteus mirabilis. Moreover, S. aureus, K. pneumoniae, Haemophilus influenzae, and Shigello flexneri can launch an offensive against the host's immune system and suppress the activity of genes responsible for the synthesis of AMP in the organism infected with them. It is time to think about the possible consequences for a person using this weapon, so that inadvertently it does not work out according to Viktor Chernomyrdin, who said «We wanted the best, but it turned out as always!», or Stanislaw Lem, who wrote that «... all of us will come to an end when the AIDS virus learns to fly.» We can bring this end closer if we manage to teach a dozen or two microbes to suppress the synthesis of AMP and thereby, our immune system.*

* *Some authors (Witkowska et al., 2008) rightly believe that the abbreviation AMP adopted in the English language literature for antimicrobial peptides is unsuccessful since it has long been used to denote adenosine monophosphate.*

For a long time, the antimicrobial effect of RJ was associated only with 10-HDA. However, in 1990, Japanese researchers (Fujiwara et al., 1990) discovered for the first time a low molecular weight peptide with antimicrobial properties in its composition.

The antimicrobial peptide of RJ was named by its discoverers as royalisin. It turned out that it consists of 51 amino acids and its molecular weight is 5532 Da. It is able to strongly inhibit the development of gram-positive microbes but does not act on gram-negative, that is, it has a narrower spectrum of action than RJ. Therefore, the search continued and later three new broad-spectrum AMPs were found in RJ. These peptides inhibit the development of both gram-positive and gram-negative bacteria, as well as yeast (Fontana et al., 2004). They were called jelleins (derived from jelly). The peculiarity of jelleins is that, unlike other known AMPs, they are very short (their molecules are built of only 8–10 amino acids) and their molecular weight ranges from 942 to 1082 Da. Due to their low mass and the presence of a positive charge at one of the ends of the molecule, they completely dissolve in water. Another important feature of jelleins is that their synthesis is carried out constitutively in the hypopharyngeal glands of workers and is in no way associated with the action of any pathogens.

The peptide composition of RJ has not yet been fully studied and the possibility of discovering new bactericidal compounds of a protein nature in it cannot be ruled out. More than 60 different peptides have been found in the mandibular glands of female workers (Santos et al., 2005) and some of them can pass into RJ.

RJ is famous, not only for its bactericidal and anti-cancer properties, but for other beneficial properties. It was one of the first to discover its insulin-like action (Dixit & Patel, 1964; Kramer et al., 1977). Its track record also includes antioxidant activity (Nagai et al., 2001; Nagai & Inoue 2004; Nagai et al., 2006; Jamnik et al. 2007), that is, the ability to prevent the occurrence of oxidative stress, which gives rise to the development of severe chronic diseases. Its estrogenic effect (Mishima et al., 2005) opens up the possibility of withdrawal from hormone therapy, fraught with an increased risk of breast cancer and coronary artery disease (Timins, 2004). This same property of RJ makes it an excellent prophylactic agent against osteoporosis, which often develops in women during menopause. It also discovered the ability to prevent unwanted complications as a result of chemotherapy and in the treatment of certain drugs, for example, paracetamol, which is harmful to the liver and kidneys (Suemaru et al., 2008; Kanbur et al., 2009). RJ is an excellent nootropic, which is an agent that improves memory, learning, and mental performance (sometimes called mental enhancers).

If we add to this the reports of antimutagenic activity (Stamenković-Radak et al., 2005), the ability to regenerate nerve cells (Hashimoto et al., 2005; Hattori et al., 2007), and activate the immune system (Vucevic et al., 2007 ; Dzopalic et al., 2011), then one might get the impression that RJ is a real panacea, a universal remedy that can solve all health-related problems at once. However, to ascribe such possibilities to RJ is like creating a new myth about it. While it is effective for many diseases and helps prevent the occurrence of many others, it is not a panacea. In some cases, for example, with Addison's disease, its use is generally contraindicated. In addition, it contains peptides-allergens with a mass of 47 and 55 kDa (Schmitzová et al., 1998), which can cause various painful effects in people with individual sensitivities, including rhinitis, eczema, and even the development of an asthmatic state. All these contraindications apply, fortunately, to a limited circle of people. Therefore, not only the healthiest and the strongest can consume RJ regularly.

Perhaps the reader has noticed that most of the publications cited above regarding the biological activity of RJ belong to the pen of Japanese researchers. Is it accidental that the scientists of a country where the average life expectancy is one of the highest in the world and which is the main importer of this product, the largest quantities of which (more than 2000 tons/year, that is, more than 60% of world production) are produced in China?

The widespread use of RJ began after the 1960s. L.N. Braines in the USSR developed a method for preserving its biological activity by adsorption. This method consists in the fact that immediately after extraction from queen cells, RJ is ground in a porcelain mortar together with lactose (97–98%) and glucose (2–3%)[9].

The material obtained by this procedure is used for the production of medicines. It has been produced under the name "Apilak" from 1961 to the present time in the form of tablets in Russia, Latvia, and Estonia. It includes calcium stearate, potato starch, and talc as binders. In Western countries, commercial food supplements are made from freeze-dried RJ. In Poland, by Apipol-Farma and BARTPOL, it is produced in capsules containing 100 mg of lyophilisate with added glucose.

---

[9] *This is probably not the best way to preserve royal jelly since the resulting product cannot be used by people with lactase deficiency, a disease in which the absorption of lactose disaccharide is impaired. Among the Eastern Slavs, about 15% of the population suffers from this disease. According to the author of this book, lactose can be successfully replaced by sorbitol.*

In addition, lyophilized RJ is included in Apivit and Apivit P (Sędecki Bartnik) preparations.

3.4. Not only a "food warehouse" but also a real "arsenal"

The study of the chemical composition of RJ was carried out by a large number of research groups, each of which used their favourite analytical techniques and methodology for this purpose. Each method has certain capabilities and limitations and because of this, the scientific reports of individual groups suffer from one-sidedness; in some, only the content of more or less complex peptides, or nucleotides and nucleic acids in RJ is reported, others are devoted to vitamins, and still others, to microelements, while in the fourth, it is written about anthropogenic pollution, and so on. There will never be any equipment that will make it possible to determine all these components simultaneously. The author of this book is also limited in research by the technical capabilities of his "own" method, gas chromatography, which provides broad but not limitless possibilities for the analysis of complex mixtures of organic compounds. For example, it is not suitable for peptide analysis. Therefore, in this section, we will only talk about relatively low molecular weight organic compounds in RJ, which form a very rich "bouquet" of substances of various classes.

For several years, we have determined the composition of both fresh RJ from different beekeeping farms in Poland and ready-made preparations purchased in pharmacies in different countries, including Russia, Estonia, Latvia, Poland, Spain, and Egypt. As a result of these analyses, we were able to identify (that is, "call by name") 169 substances in RJ. Another 23 compounds were registered, but we could not identify them, so future researchers still have a lot to work on.

All substances discovered in RJ can be conditionally divided into three groups: volatile components, non-volatile low-polarity, and non-volatile highly polar ones. Each of these groups was determined using different extraction methods: for volatile compounds, we used solid-phase microextraction (SPME), and for the other two groups, extraction with solvents of different polarities (Isidorov et al., 2012).

Figure 3.2. Breeding frame with plastic bowls on corks.
From these bowls we selected fresh royal jelly for analysis. Foto: S. Bakier.

3.4.1. A volatile defensive weapon in royal jelly: against whom is it directed?

In the volatile fraction, the chromatogram of which is shown in Fig. 3.3, we were able to identify 25 compounds (Isidorov & Bakier, 2011). Some of them are metabolites, characteristic of all living organisms, such as ethanol, acetone, and ethyl acetate (probably, not everyone knows that these three substances, along with isoprene, are the main volatile components in air exhaled by a healthy and a non-drinking person). Another group is formed by substances with a strong antimicrobial effect: phenol, methyl benzoate, methyl salicylate, *o*-guaiacol, and benzoic acid. These bactericidal compounds account for approximately 27.5% of the volatile excretion of RJ. About the same amount is accounted for by two more compounds: 2-heptanone and octanoic acid. These two compounds deserve special attention.

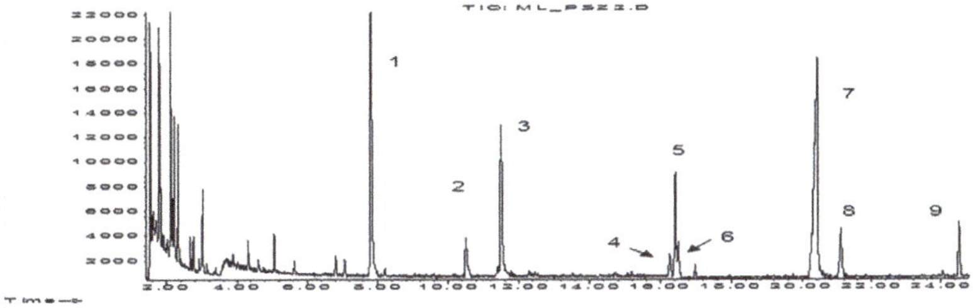

Figure 3.3. Chromatogram of volatiles of fresh royal jelly. Main components: 1 - 2-heptanone; 2 - benzaldehyde; 3 - phenol; 4 - o-guaiacol; 5 - 2-nonanone; 6 - methyl benzoate; 7 - octanoic (caprylic) acid; 8 - methyl salicylate; 9 - benzoic acid.

The first of these, 2-heptatone, as mentioned in the previous section, is produced by the maxillary glands of bees and serves as an anxiety pheromone and exhibits repellent properties (Kerr et al., 1974). 2-Heptanone is not only secreted by our honey bees *A. mellifera* but, as was found later, also by the eastern bee *A. cerana* (Naik et al., 1992) and the dwarf bee *A. florea* (Naik et al., 2002). The authors of these works showed that in both species it plays the role of a repellent.

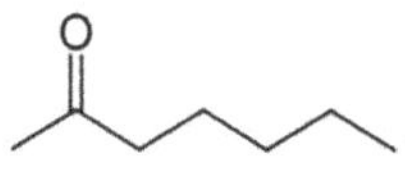

1

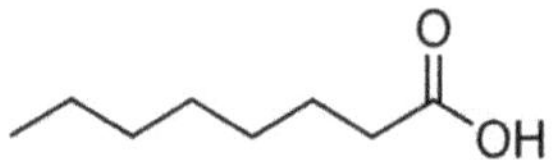

2

Figure 3.4. Chemical formulas of volatile components of royal jelly with repellent properties. 1 – 2-heptanone, 2 – octanoic acid.

The production of 2-heptanone by the maxillary glands of *A. mellifera* worker bees increases with age, reaching a maximum in foraging bees (Boch & Shearer, 1967). It is also high among the bee guards, which, apparently, use it to scare off thief bees. Previously, 2-heptanone (as well as its homologue 2-nonanone) was not detected in RJ, but as can be seen in Fig. 3.2, it is the main volatile component. This means that not only in the oldest flight bees but also in the glands of the youngest nurse bees, this ketone is produced in sufficiently large quantities.

The second substance, octanoic acid, which also possesses repellent properties, has long been discovered both in the secretion of the maxillary glands of workers and in RJ (Boch et al., 1979). It cannot be ruled out that the repelling effect of 2-heptanone and octanoic acid is mutually reinforced, that is a phenomenon called *synergism* takes place.

The question is against whom the deterrent effect of both of these main volatile components of RJ is directed. Most likely, against some parasites, for which the larvae are a welcome delicacy, but what kind of parasites is unclear. It would seem that the answer lies on the surface, against the *Varroa destructor* mite. However, it can hardly be said with certainty that it is directed against *Varroa*, since this pest has become apparent to us relatively recently. After all, bees of the species *A. mellifera*, in the course of their evolution, could not develop in advance a means of defence against the impending danger.

I suggest that initially it could be directed against other enemies of the honey bee, which, like *V. destructor*, develop on larvae, for example, ticks from the genus

Erythraeida (*Leptus* sp.) and *Pyemotida* (*Pyemotes* sp.). It can be assumed that with the passage of time between these types of parasites and the honey bee, a kind of balance has developed, parasitizing and developing in the family, they, nevertheless, do not lead to its complete death. *Acarapis* ticks also belong to the hemolymph lovers. These are *A. externus* (localized at the base of the head from the side of the abdomen), *A. dorsalis* (settles between the second and third segments of the body), and *A. vagans* (digs into the body of the bee near the base of the first pair of wings near the first segment of the body). However, the biology of these pests is still poorly understood.

One way or another, in laboratory and field studies, it has been reliably shown that octanoic acid repels *V. destructor* (Nazzi et al., 2009). Therefore, this may be the reason for the different degrees of damage to larvae of different castes by this pest (Calderone et al., 2002), which occurs most often in drones, less often in workers[10], and very rarely in queens.

The female *V. destructor* mite enters the cells with the larvae just before the worker bees seal them with a wax lid. In the case of queen cells, sealing occurs after 4.5 days, cells of workers after 5.5 days, and drone cells, only on the seventh day after the appearance of the larva. At the same time, the sealed queen cells contain large quantities of RJ, and from the cells with the worker and drone brood, its smell has long disappeared and does not bother the pest at all!

The higher attractiveness of the drone brood for the *Varroa* female in comparison with the worker larvae can be explained by different reasons or their complex. One of them is the larger size of the larva, prepupae, and pupae of the drone; the conditions for development of the tick on them are better. Another likely reason is the longer developmental period of the drone from the larval stage to imago; if in the case of worker bees this period is 12, then the drones have 14 days. Therefore, during the full development cycle of a drone in a sealed cell, ticks of three successive generations can appear in it. But in any case, the *Varroa* must somehow "identify" the drone cell and distinguish it from the worker cells. Specific volatile compounds secreted by drone larvae act as kairomones for the parasite, which can serve as a reference point for it.

---

10 *According to Prof. E. Wilde (Poland), cells with workers' larvae are colonized by a mite 14.5 times less frequently than drone cells.*

Kairomones are substances released by the body into the environment and affecting other organisms. Kairomones include odorous secretions that attract parasites and predators. They are dangerous for individuals that emit them but they contribute to the regulation of population size. Thus, they can be beneficial for the species as a whole.

If these kairomones are discovered, they can be used to combat varroatosis, as bait for the mite placed in a special trap.

In summary, it can be argued that more than 50% of the volatile secretions of fresh RJ are part of the arsenal of defensive agents, are of a protective nature, and closer attention to them can help in the search for remedies against honey bee pests, microorganisms, and parasites.

3.4.2. Why is royal jelly "too tough" for microbes?

Low-polarity compounds were extracted from fresh RJ with diethyl ether. Here are some of the results of just one series of studies, in which 17 samples of RJ obtained in Poland from bees of the same race (*A. mellifera carnica*) were analyzed. This fraction is formed by 85 compounds, 22 of which were discovered by us for the first time. Of these, 47 compounds turned out to be hydroxy acids with an even number of carbon atoms in the molecule: 8, 10, 12, and 14 (Table 3.2). Such a set of acids is unique, as they are not found in any other natural object.

Table 3.2. Chemical compounds of royal jelly

Group of compound	Representatives of the group
Saturated monohydroxy acids	2-, 3-, 7- and 8-Hydrooxyoctanoic acids, 3-, 9-, and 10-hydroxydecanoic acids, 3-, 11- and 12-hydroxydodecanoic acids, 3-methyl-3-hydroxyglutaric acid; 3-hydroxydodecanodioic acid
Saturated dihydroxy acids	3,9-, 3,10-, 3,11-and 8,9-Dihydroxydecanoic, 9,10-, 10,11-, 11,12-, and 3,12-dihydroxydocecanoic, 3,13-dihydroxytetradecanoic
Unsaturated hydroxy acids	7-, and 8-hydroxy-2-octenoic acids, 8-hydroxy-2-decenoic acid, (Z)-, and (E)-9-hydroxy2-decenoic acids, (E)-10-hydroxy-2-decenoic acid (10-HDA); 11-, and 12-hydroxy-2-dodecenoic, 9,10-dihydroxy-2-dodecenoic, 13-hydroxy-2-tetradecenoic, 11,12-dihydroxy-2-tetradecenoic
Oxo acids	2-Oxooctanoic, 9-oxo-2-decenoic (9-ODA)
Dicarboxylic acids	Succinic, octanedioic (suberic), decadioic (sebacic), dodecanodioic, 2-octeno-1,8-dioic, 2-deceno-1,10-dioic, 2-dodecene-1,12-dioic (traumatic)
Aliphatic acids	Hexadecanoic (palmitic), octadecanoic (stearic), iscosanoic, tetracosanoic, oleinic
Aromatics	Benzoic acid, 4-hydroxybenzoic acid (parabene), 4-hydroxydihydrocin-namic, p-coumaric acid, caffeic acid, 4-hydroxy-3-methoxyphenyl ethanol (HVA), methyl-4-hydroxybenzoic acid (methyl parabene, HOB), pyrocatechol, hydroquinone, nicotinic acid
Alkanes and alkenes	n-Pentacosane, n-heptacosane, n-nonacosane, n-hentriacontane, 7-, and 9-hentriacontene, 9-tritriacontene
Other compounds	Lactic acid, 2,3-butanediol (two isomers), glycerol, 1-hexadecanil, 1-tetracosanol, cholesterol

Judging from the literature, in addition to the above compounds, RJ also contains 10-acetoxydecanoic, 10-acetoxy-2-decenoic, and 11-oxododecanoic acid (Melliou & Chinou, 2005), as well as 12-hydroxytetradecanoic acid (Drijfhout et al., 2005). However, despite all efforts, we failed to find these substances in any of the 17 samples.

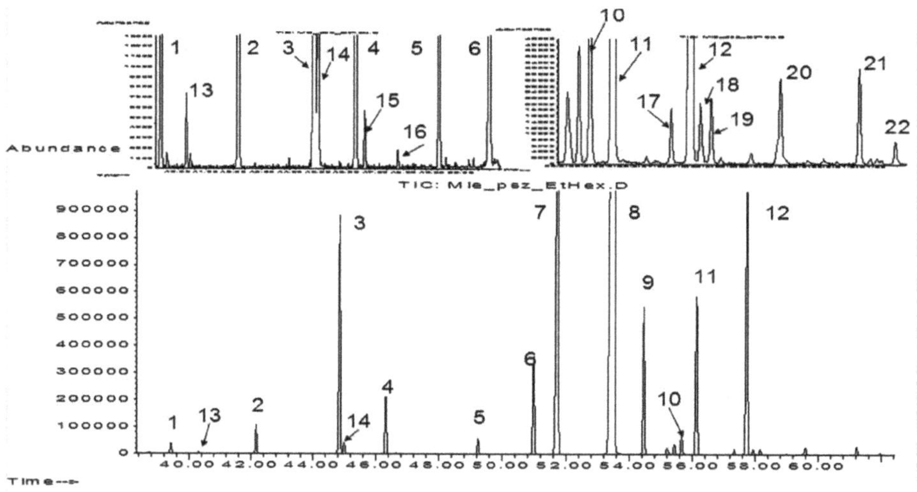

Figure 3.5. Chromatogram of the acidic part of the ether extract of fresh royal jelly.
Main peaks: 1 - 3-hydroxyoctanoic acid; 2 - 7-hydroxyoctanoic acid; 3 - 8-hydroxyoctanoic acid; 4 - 3-hydroxydecanoic acid; 5 - 9-hydroxydecanoic acid; 6 - 9-hydroxy-2-gesoic acid (9-HDA); 7 - 10-hydroxydecanoic acid (10-HDAA); 8-10-hydroxy-2-decenoic acid (10-HDA); 9 - sebacic acid; 10 - 3,9-dihydroxydecanoic acid; 11 - 2-decene-1,10-dicarboxylic acid; 12 - 3,10-dihydroxidecanoic acid.
Left insert: 13 - methylparaben; 14 - 4-hydroxybenzoic acid (paraben); 15 - 8-hydroxy-2-octenoic acid; 16 - suberic acid.
Right inset: 17 - 11-hydroxy-2-dodecenoic acid; 18 - 12-hydroxydodecanoic acid; 19 - 8,9-dihydroxydecanoic acid; 20 - 9,10-dihydroxy-2-dodecenoic acid; 21 - 3,11-dihydroxydodecanoic acid; 22 - 10,11-dihydroxydodecanoic acid.

As seen in Fig. 3.4, the contents of different acids are different. Many of them are present in small, so-called trace, quantities. However, their role in the formation of the biological properties of RJ may not be negligible.

In full agreement with the data of other authors (Lercker et al., 1981; Melliou & Chinou, 2005), the main components in all 17 samples were 10-HDA (50.3–66.7%), 10-hydroxydecanoic acid (13.0–17.6%), 3,10-dihydroxydecanoic acid (4.4–7.8%) and 2-decene-1,10-dicarboxylic acid (3.1–5.6%), which we discovered for the first time in RJ.

Notably, ether extracts contain many chemicals with a similar structure. For example, a group of unsaturated hydroxyacids forms 12 compounds, which have a double bond in the second position, counting from the "acid" end of the molecule, and they themselves differ only in the number of $–CH_2–$ units in the molecule (chemists call such compounds homologues). Usually, substances with a similar structure exhibit similar properties, including biological activity.

The ability to suppress the development of a wide range of pathogenic bacteria was discovered in RJ more than 80 years ago (McCleskey & Melampy, 1938) but the nature of this phenomenon was completely unknown for a long time. Soon after the discovery of 10-HDA in this product, studies began on the biological activity of this acid, including its antibacterial action. An American research group reported that this hydroxy acid in tests on a representative group of bacteria and fungi (carried out by Glaxo Ltd.) showed no activity or it was weak (Barker et al., 1959). However, at practically the same time, another research group demonstrated the high activity of 10-HDA against many bacteria and fungi (Blum et al., 1959). Quite reliable results were obtained in both cases. The reason for this striking discrepancy has not been discussed anywhere in the scientific literature, although it is instructive. Barker et al. (1959), apparently, did not pay attention to the message in the work cited above (McCleskey and Melampy, 1939) that RJ killed the tested microbes within a few minutes at its natural acidity (pH 4.6) but in a neutral environment (pH 7.0) it took two days. Blum et al. (1959) showed this in their tests carried out under slightly acidic conditions and thus the bactericidal effect of 10-HDA was observed.

Until recently, there was an idea of such activity of only one of the acids of RJ. It is 10-HDA that is credited with such important properties as antimicrobial and anti-cancer effects. But, perhaps, its homologues, which are also contained in RJ are even more active. Therefore, it is not so important that they are contained within it in incomparably smaller quantities.

I will give an example of the meaning of "small" impurities from another area of my scientific interests, from atmospheric chemistry. It is known that the main gasses of the atmosphere (N2 and O2) do not affect the change in the magnitude of the greenhouse effect. It entirely depends on changes in the concentration of "small" gases, mainly CO2, CH4, and N2O. The content of methane CH4 and N2O (this oxide is emitted into the air by our cars and our fields, abundantly fertilized with nitrates) in the atmosphere is hundreds of times less than carbon dioxide. However, the global warming potential of methane is 25 times and N2O is 298 times that of CO2. Therefore, the contribution to the greenhouse effect of each of these three gasses is comparable. Conversely, the growth rate of the concentrations of methane and N2O in the atmosphere is higher than the growth rate of the carbon dioxide content. Consequently, limiting CO2 emissions, while increasing the emissions of other «minor» greenhouse gasses, to put it mildly, is unreasonable.

Returning to our acids, I note that not only 10-HDA was reported to have bactericidal properties. They are inherent in sebacic and 3-hydroxydecanoic acids, the content of which in RJ is not low; in all 17 samples, it averaged 3.3 ± 0.6% and 1.2 ± 0.2%, respectively.

3-Hydroxydecanoic acid was first found in wildlife in the secretions of the glands of the South American ants, Atta sexdens (Schildknecht & Koob, 1971). This compound, called by the authors myrmicacin (myrmica is a genus of ants), is used by ants to prevent the germination and rotting of seeds they bring to the nest, that is, it acts as an herbicide and aseptic. The antimicrobial properties of myrmicacin, especially active against pathogenic gram-positive bacteria (staphylococci, streptococci), were first demonstrated by Japanese researchers (Iizuka et al., 1979).

In antimicrobial activity, other RJ acids related to myrmicacin include 3-hydroxyoctanoic and 3-hydroxydodecanoic. RJ always contains 2-dodecene-1,12-dioic acid, otherwise called traumatic acid. This name was given to it due to the fact that it appears in plant tissues after damage and stimulates wound healing. In addition, like other unsaturated acids, it exhibits antimicrobial action. The content of traumatic acid in RJ is small, almost at the level of trace amounts. However, it contains other homologues very similar to traumatic acid: 2-octene-1,8-dioic and

2-decene-1,10-dioic, the formulas of which and the average contents in the ether extracts from 17 samples are given below.

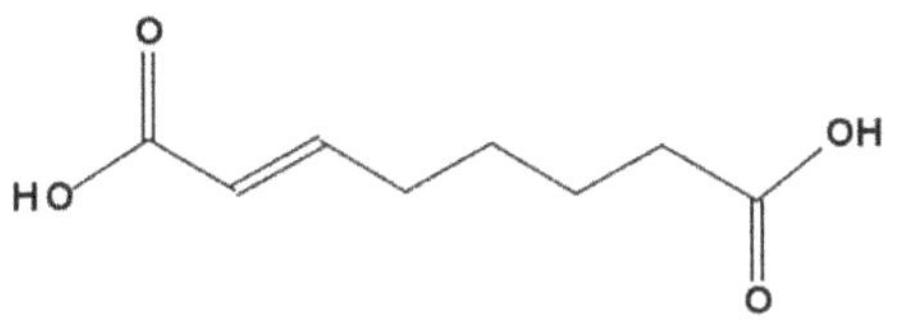

1

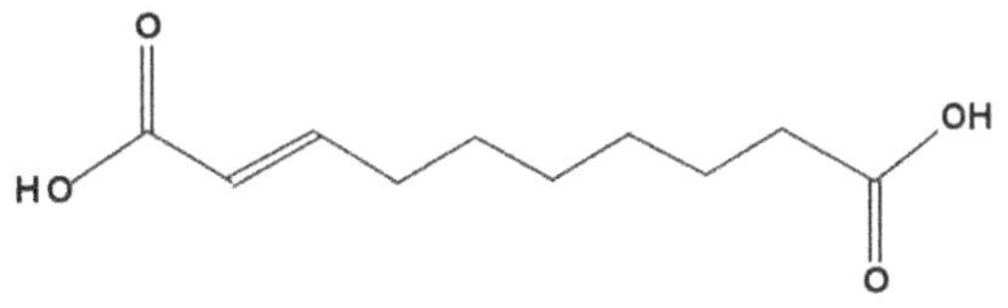

2

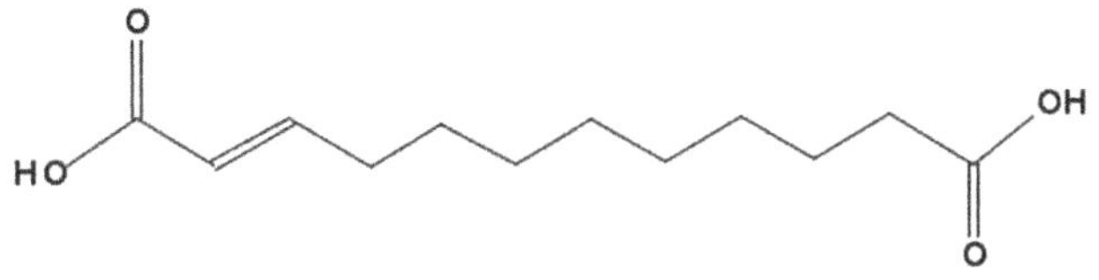

3

Figure 3.6. Unsaturated dioic acid in royal jelly. 1 – 2-octene-1,8-dioic acid (0.06±0.04%), 2 – 2-decene-1, 10- dioic acid(4.1±0.8%), 3 – 2-dodecene-1,12-dioic (traumatic) acid.

It cannot be ruled out that the healing effect of RJ noted by Japanese scientists is associated with the complex of these acids (Fujii et al., 1990).

Recently, we have carried out studies of the antimicrobial effect of some of these components of RJ and confirmed their high activity against the tested gram-positive bacteria, but a weaker effect on gram-negative ones (Isidorov et al., 2018). The result of these tests were carried out under slightly acidic (pH 5.1) conditions (Table 3.3).

Table 3.3. Antimicrobial activity of selected RJ acids and crude RJ

Bacterium	Minimum inhibitory concentration, MIC (mMol)								
	$2\text{-}HOC_8$	$8\text{-}HOC_8$	9-HDDA	10-HDAA	12-HDAA	10-HDA	2-DecDA	2DDecDA	RJ, µg/ml
Gram-positive bacteria									
Staphylococcus aureus	0.78	1.49	0.048	3.12	0.048	0.098	0.78	0.39	62.5
Bacillus subtilis	0.78	3.12	0.048	0.78	0.098	0.098	0.39	0.048	31.2
Bacillus cereus ATCC 14597	3.12	6.25	0.048	0.039	0.039	0.78	1.195	0.78	62.5
Bacillus cereus F4810/72	0.78	3.12	0.048	0.039	0.195	0.39	0.048	0.78	62.5
Bacillus thuringiensis	1.56	6.25	0.048	0.78	0.39	0.39	0,39	0.39	125.0
Gram-negative bacteria									
Pseudomonas aeruginosa	6.25	12.50	12.50	12.50	6.25	6.25	1.25	2.50	250.0
Escerichia coli	6.25	12.50	12.50	12.50	6.25	3.125	2.50	5.00	250.0

$2\text{-}HOC_8$ - 2-hydroxyoctanoic acid, $8\text{-}HOC_8$ - 8-hydroxyoctanoic acid, 9-HDDA -9-hydroxydecanoic acid, 10-HDDA -10; -hydroxydecanoic acid, 10-HDA- 10-hydroxy-2-decenoic acid, 2-DecDA - 2-decene-1,10-dioic acid; 2DDecDA -2-dodecene-1,2-dioic (traumatic) acid.

As you can see, all tested RJ acids inhibited bacterial growth and the main one, 10-HDA, was not the most active in this regard.

It can be assumed that it is the presence in RJ of many compounds with a similar structure and/or similar biological activity that is the main reason why microbes (how many there are) for many millions of years have not been able to find a "key" to it. We have to admit that the natural pharmacy of bees is incomparably more effective than the entire modern chemical and pharmaceutical industry and we have a lot to learn from them.

In recent decades, the search for biologically active compounds of natural origin has been carried out on a large scale with the aim of their further use in medicine, cosmetology, and as preservatives in the food industry. However, the dominant trend in the production of medicines containing individual artificially synthesized natural compounds (from those found as a result of such screening) with the addition of more or less inert fillers raises doubts. In living nature, they always play their role in the accompaniment of a whole orchestra, consisting of other substances, but «singing a cappella» often does not give the same effect.

At the current level of development of the chemical and pharmaceutical industry, the production of the same 10-HDA is not particularly difficult. But there is no guarantee that even against this powerful antiseptic, microbes will not find remedies, as has been the case with antibiotics. If this does happen, then the consequences for both humans and bees, with which they come into contact, are unknown. According to Stanislaw Lem (2000), "learning from a master, what biology should have been, should not be replaced by peeping at it».

The chemical nature of the antioxidant action of RJ is not entirely clear to me. According to Japanese researchers, in terms of these properties, it is inferior to propolis but noticeably superior to any types of monofloral honey (Nagai et al., 2001). Hydroxy acids, even unsaturated ones, based on general chemical considerations, should not have a high antioxidant potential. The content of vitamin E, a well-known antioxidant, in RJ is also not very high. Some sources say it contains phenolic compounds. Simple phenols and phenol carboxylic acids, in fact, actively bind free radicals. But of all the compounds of this group, we found only paraben (4-hydroxybenzoic acid) in RJ and even then, in very low concentrations ($0.2 \pm 0.1\%$ of the ether extract). Trace amounts of pyrocatechol and hydroquinone, as well as *p*-coumaric and caffeic acids (<0.01%), which were not found in all samples, cannot significantly affect the antioxidant properties of RJ. Moreover, I am inclined to consider *p*-coumaric and caffeic acids as random impurities; they belong to the main components of propolis, with which the bees polish the cells, and can be extracted with RJ from the walls of these cells.

We have to assume that the antioxidant effect of RJ is mainly associated not with the extractable components, but with more complex compounds, most likely of a protein nature or even with the products of their enzymatic transformations. The proteins contained in it are easily broken down by animal digestive enzymes, pepsin, and trypsin, as well as papain, "plant pepsin". According to Japanese scientists, the resulting products have very high antioxidant activity (Nagai et al., 2006). These authors concluded that RJ is a health food because it prevents the onset of diseases based on oxidative stress.

The above components of the ether extracts of RJ include a group of five compounds of known origin but it is not known where, how, and why they got into its composition. These are *para*-methoxybenzoic acid (HOB), 4-hydroxy-3-methoxyphenylethanol (HVA), 9-oxo-2-decenoic acid (9-ODA), and two isomers of 9-hydroxy-2-decenoic acid, (*Z*)-9-HDA and (*E*)-9-HDA. These are all components of the queen bee pheromone, produced in her maxillary glands (their formulas are given in the second chapter on Fig. 2.2). So how did these substances get into the RJ, in the production of which the bee queen does not participate?

German researchers have found that each of them begins to be synthesized in the glands of the queen at a certain time (Engels et al., 1997). To clarify this dynamic, they removed the glands from the heads of queens of different ages and subjected them to extraction (in each series of the experiment, which lasted three months, 20 bees were used). To minimize as many random variations as possible in the composition of the glands, the scientists specially raised sisters for their experiments, that is, bees descended from the same queen. It turned out that the synthesis of 9-HDA and 9-ODA begins in the glands of young bees the day after leaving the queen cell and their content reaches a maximum by 8–10 days of life. Even later, HOV and HVA appear in the glands; the first on the 15th day and the second on the 90th day of their life. According to the authors, these two substances, especially HVA, are characteristic of a mature and dominant queen. As discussed in the previous section, HVA acts as a pacifying agent on young worker bees by blocking their queen rejection response (Vergoz et al., 2007).

Our research identified a set of five components that form the pheromone of the queen in only four out of 17 RJ samples, which was 24% of samples. HOB and HVA were found in 53 and 35% of the samples and only one of the (*Z*)-9-HDA isomers was present in all samples and in the highest amounts (1.7 ± 0.4%). The content of the second isomer, (*E*)-9-HDA, was found in eight samples but did not exceed 0.02%. The concentration of 9-ODA was at the same level. This again raises the question of where and how these substances got into RJ and why were they not found in all samples.

It seems incredible to me to assume that they enter the RJ from the body of the egg-laying queen or from its maxillary glands. In this case, many other characteristic substances would have to be present in the RJ based on the composition of extracts from these glands, of which Engels et al. (1997) found over 120 different compounds. Then the assumption suggests that the components of the pheromone of the queen are formed in the maxillary glands of some workers, mixed with the secretions of the pharyngeal glands, and then enter RJ. Maybe this occurs in "anatomical" laying workers that are just about ready to become physiological laying workers and start laying eggs. After all, it is for them that the production of these compounds is useful as a disguise under the queen. In the previous chapter, I gave information that some laying workers even manage to acquire their own retinue. However, this is just an assumption and finding out the true reason for the presence of queen pheromone components in RJ requires the participation of specialists in the field of bee biology.

As previously discussed, in RJ dehydrated by lyophilisation, the sugar content reaches 30% but this product does not differ in sweetness. It tastes sour and slightly pungent; the taste of sugars is interrupted by acids, although there are noticeably fewer of them. Sugars are strongly polar compounds and for their extraction we used a polar solvent, methanol. In this fraction, 82 compounds were found but most of them were contained in small quantities. About 79% accounted for only five substances: simple sugars, represented by fructose, glucose, and galactose and disaccharides, sucrose and trehalose. In addition to these, RJ contained 16 other disaccharides and 10 tri- and tetrasaccharides. Together they constituted approximately 6% of the fraction.

Simple sugars and most disaccharides are present in various cyclic forms. For example, three isomers were discovered in RJ for fructose and glucose. Fructose is represented by two five-membered (α- and β-fructofuranose) and one six-membered (β-fructopyranose), and glucose by two six-membered (α- and β-glucopyranose) and one five-membered (β-glucofuranose) cycles. In aqueous solutions of sugars, an equilibrium is established between all ring-shaped structures (called anomers). Transitions between anomers are thought to be through a non-cyclic form, although this is usually not detected. In our studies, we were also unable to register the presence of α-glucofuranose.

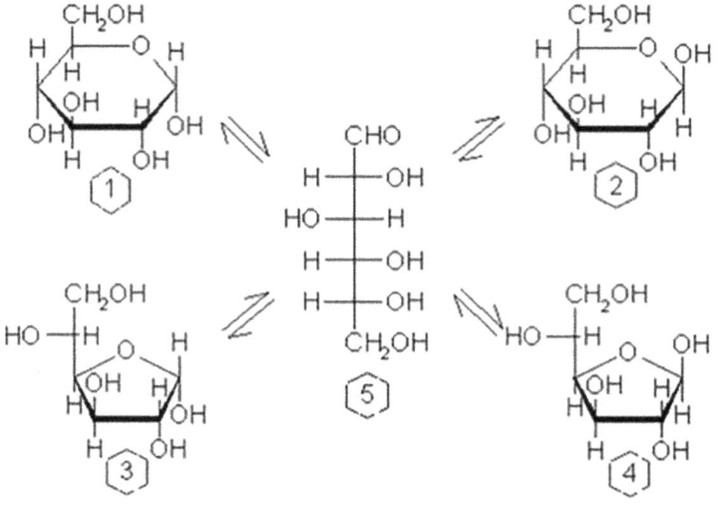

Figure 3.7. Mutual transformation (mutarotation) of glucose. 1 - α-glucopyranose, 2 - β-glucopyranose, 3 - α-glucofuranose, 4 - β-glucofuranose, 5 - D-glucose.

Such subtle differences in the composition of sugars can be seen only when using gas chromatography, which has the highest separation power (Lercker et al., 1986; Isidorov et al., 2009), but they elude attention when using liquid chromatography (Sesta, 2006). This makes it difficult to make any assumptions about the role of different forms of sugars in biological processes. Notably, their content is different from the equilibrium content in an aqueous solution. For example, in water there are only two forms of glucose, α- and β-glucopyranose, in a percentage ratio of 35:65. In RJ, they were present in approximately equal amounts, although their total contents varied greatly from sample to sample. Other authors also noted high variability in the content of all sugars. For example, according to Sesta (2006), the concentration of fructose in samples of RJ changed threefold (2.3–6.9 g/100 g), glucose changed 2.2 times over (3.7–8.2 g/100 g), and sucrose and maltose changed more than 20 times over. Against the background of such fluctuations, the reason (cited at the beginning of this chapter) that the ratio of glucose and fructose are *somewhat different* in the feed of the larvae of queen bees and workers does not look serious without being supported by the statistical processing of research results.

With a certain degree of certainty, we can say that glucose, fructose, sucrose, and galactose in RJ are of plant origin. In contrast, trehalose is not found in nectar or pollen. This non-reducing disaccharide is formed in the body of the bee from two glucose molecules.

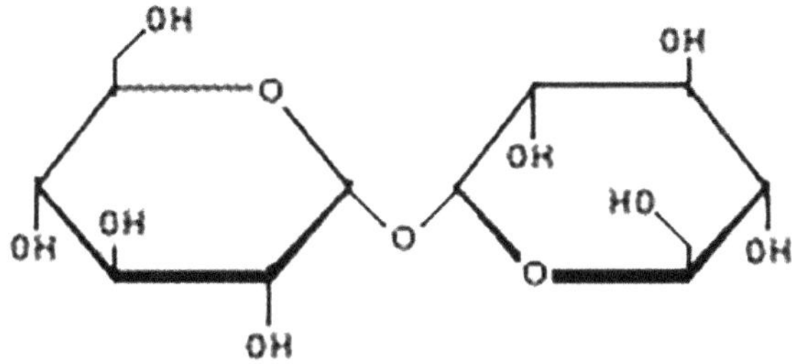

Figure 3.8. Trehalose formula.

Trehalose is the main sugar of the bee's hemolymph, which performs a transport function. It transfers glucose to the insect tissues that need it. If we talk about the properties of trehalose, then we must name its ability to retain large quantities of water molecules, which is why trehalose is produced only in plants from arid and desert regions. Its antioxidant capacity has also been reported, but I could not find more detailed information on this subject.

Other compounds, besides sugars, are also present in this fraction of RJ. There are compounds related to sugars, the products of their oxidation (sugar acids), substitution (glycosides), and reduction (alcohols sorbitol, mannitol and others). Enzymatic reduction of glucose also leads to the formation of cyclic alcohols, known as inositols, which are included in the complex of vitamins of group B. RJ contains *myo*-inositol in the largest quantities. This is an important compound since it is involved in the restoration of the structure of nerve tissues and in protecting against damage to brain cells.

Among other biologically active components in RJ, nitrogen-containing compounds should be mentioned, amino sugars, and nucleosides (adenosine and guanosine). Amino derivatives of glucose and galactose are part of chitin, of which the outer covers of the daily molting bee larva are composed. Therefore, it cannot be ruled out that N-acetyl glucosamine is not produced by the glands of female workers but is secreted into the RJ by the larvae themselves.

Adenosine and guanosine are part of nucleic acids. Adenosine is also an integral part of ATP, the main energetic substance of all living organisms. In addition, it plays an independent role in the transmission of cellular signals. We found adenosine in all analyzed samples but its content varied 20 times. Even larger fluctuations in adenosine concentrations, from about 6 to 2060 μg/g, were noted in 45 samples of RJ by Chinese researchers (Xue et al., 2009).

This concludes my review of the chemical composition of RJ. It shows that the head (maxillary and pharyngeal) glands of the honey bee are a real biochemical factory. The overwhelming majority of the components of RJ listed in this section are produced at this factory. The exception, and even then, not 100%, may be sugars such as glucose, fructose, and sucrose. They can be exclusively of plant origin, as a sweet tribute to plants. Neither 10-HDA nor other hydroxy acids like it enter the hive from the outside, as well as peptides specific to RJ. The initial raw material for the synthesis of peptides, acids, and many other compounds are organic matter of plant pollen but little of its composition goes directly into RJ. Plant proteins are decomposed by bee enzymes into amino acids and new ones are built from these "bricks" according to the hereditary program recorded in the genes. All unique hydroxy-, oxo-, and dicarboxylic acids are synthesized in the glands from a single precursor, also supplied with pollen, from stearic (octadecanoic) acid (Pletner et al., 1996). As a result of all these complex processes, a matter of amazing quality is obtained, called RJ.

3.5. Useful "impurities" of royal jelly

The mysteries of RJ, produced by worker bees are not limited to the presence of queen pheromone components. We were greatly disturbed by the incomprehensible discrepancy between our own and published analysis results. These discrepancies relate to the content of free (that is, not included in the composition of proteins) amino acids. When starting our research, we expected to find them in RJ, since Italian researchers (Boselli et al., 2003) reported that they were able to identify nine free amino acids. Indeed, in one of the Apilak samples produced in Latvia, we found eight of them in a free state (Isidorov et al., 2009). However, it was only found in one of six other samples of Apilak from the same company. As it turned out in the pharmaceutical preparations of RJ, purchased in Russia, Poland, Spain, and Egypt, they were also not found in any of the more than 30 samples of Polish fresh RJ. However, all samples contained one amino acid, proline, which is a special compound. Chinese researchers (Liming et al., 2009) discovered 25 more free amino acids in addition to proline in domestic fresh RJ and in rather large quantities with an average of 9.21 mg per 1 g of RJ.

We cannot blame our technique since it makes it possible to register amino acids at a much lower level of their content than was given in the works of Italian and Chinese authors. The mystery was explained when we decided to see what we could obtain from the larvae themselves. We carefully removed the larvae (3–4 days old) from the queen cells (in other experiments, from plastic bowls for raising queens), washed them with methanol, and then rubbed them in a fresh portion of this alcohol. The homogenate was centrifuged, and part of the supernatant was used for analysis. Notably, we did not expect any special "sensations", but, to our great amazement, a "forest of peaks" of 22 amino acids grew on the chromatogram, including the most valuable.

Careless handling of larvae floating in RJ, even a simple injury to them, will lead to the appearance of free amino acids in it. That is, formally speaking, it is contaminated with foreign substances. This, probably, is the reason for our "discovery" of these compounds in one single series of the Latvian Apilak. Obviously, some careless worker did not bother to carefully remove the larvae from the bowls and the royal jelly he collected was contaminated. According to the Chinese authors, free amino acids were found in all ten samples received from various beekeeping farms in the country. It is not known how the RJ was selected in these farms, but one can guess that it was not too careful and part of the larvae was damaged, which led to its pollution with free amino acids.

This is a paradox as the nutritional value of the product from this kind of pollution only increases. Not only free amino acids (easily digestible) but also other valuable contaminants contained in the homogenate of larvae pass into RJ. Another undervalued product, which is a homogenate of honey bee larvae, will be discussed in the next Chapter.

Chapter 4

Drone homogenate: An invaluable beekeeping product

In one of the literary sources devoted to RJ, I happened to read that some unscrupulous manufacturers fake this expensive and demanded product, adding extracts from drone larvae to it. This pomace in colour and consistency resemble RJ and are completely mixed with it, practically not changing the taste (in both cases, it is slightly sour). Therefore, such a fake is difficult to detect. The paradox, I note again, is that unlike other cases of falsification (for example, by adding sugar syrup to honey), this does not lead to a decrease in the consumer properties of the main product. However, this circumstance does not excuse these deceivers but it gives reason to think about the possibility of producing a new product with properties different from both pure RJ and drone brood homogenate (sometimes called drone milk or larval milk). Moreover, it combines, and even enhances, nutritional and medicinal properties in comparison with individual ingredients due to synergy; in this case, the irreducibility of the properties of a new product to a simple sum of properties of its components.

The main feature that determines the nutritional value of bee brood homogenate is the high content of free amino acids, including essential ones. In this case, *essential* means only that they are not synthesized in the human body and their lack leads to serious disturbances in metabolic processes. They must constantly enter our body from the outside and there are not many food products that contain them in a complete set. In the absence of at least one of these acids, the synthesis of proteins necessary for the body becomes impossible and in this case food is used only as a source of energy or turns into fatty deposits.

Nine α-amino acids are considered essential, including valine, leucine, isoleucine, lysine, methionine, threonine, tryptophan, phenylalanine, and histidine. In addition, three more amino acids must be included in the children's diet, arginine, tyrosine, and cysteine, as essential. The rest of the amino acids can be formed in the body of an adult and child from other amino acids. Many amino acids are known in nature but most proteins contain only 18 of them, therefore they are called proteinogenic. Proteins are conventionally divided into complete, containing all essential amino acids, and non-complete.

Our studies of the composition of homogenates of queens and drone larvae did not reveal any noticeable differences in the composition of free amino acids; in both, we found 25 compounds each (Isidorov et al., 2012). They can be divided into four groups: essential proteinogenic, nonessential proteinogenic, non-proteinogenic, and "intermediate" amino acids. The latter are involved in the biosynthesis of protein acids as their precursors or are formed from them in the course of metabolism. For example, the non-proteinogenic amino acid homoserine is synthesized in plants from aspartic acid and serves as a precursor of essential proteinogenic acids, isoleucine, threonine, and methionine.

Table 4.1. Free amino acids in drone homogenate (Isidorov et al., 2012).

Proteinogenic amino acids		Intermediate amino acids	Non-proteinogenic amino acids
essential	non-essential		
valine, leucine, isoleucine, lysine, methionine, threonine, tryptophan, phenylalanine, histidine	glycine, alanine, serine, proline, asparagine, glutamine, tyrosine, aspartic acid	homoserine, sarcosine, 4-hydroxyproline, 5-oxoproline	β-alanine, β-aminoisobutyric acid, γ-aminobutyric acid

Of the protein α-amino acids, two were missing in our samples: arginine and cysteine. However, this does not mean that they are absent in all larvae homogenate. The simple method we used did not reliably detect these compounds. They were found using a different technique in the larvae of queens, drones, and workers in samples obtained from four different regions of Russia (Lazaryan, 2002). As for the rest, our results and those presented in this publication are in good agreement with each other, both in terms of the qualitative and quantitative composition of amino acids. Most of all, free nonessential and essential acids (4.1±0.3 and 6.1±0.3%, respectively) were in homogenates of queen larvae. In methanol extracts from drone larvae, they were contained in amounts of 5.1±0.4 and 3.7±0.3%, respectively.

In addition to proteinogenic compounds, homogenized bee larvae contain compounds that are not part of proteins but, nevertheless, play an important role in living organisms. For example, β-alanine is a precursor to carnosine, a dipeptide of muscle and brain tissue. The most important neurotransmitter of the human central nervous system is γ-aminobutyric acid, which also has a nootropic effect. Due to this property, it is prescribed for attention deficit disorder, which is often observed in children. True, this amino acid is present in the larval homogenate in small (trace) amounts.

Of course, amino acids of the larval homogenates are contained in them, not only in free form, but most of them are in proteins. Since these proteins contain all the essential amino acids, they are considered complete. The total peptide content in the raw homogenate is approximately 38% and in the freeze-dried homogenate, it is 52% (Lazaryan et al., 2003). That is, there are even more of them than in RJ. The table below provides a comparative description of the total composition of these products (in freeze-dried form) obtained from one of the apiaries in north-eastern Poland (Isidorov et al., 2016).

Table 4.2. Comparison of the general characteristics of the composition of drone homogenate and royal jelly.

Characteristic	Drone homogenate	Royal jelly
Water, %	3.0±1.0	2.7±0.9
Proteine, %	32.0±2.9	34.0±9.9
Lipids, %	24.2±1.0	13.5±7.8
Carbohydrate, %	38.9	26.8±5.8
Ash, %	2.7±0.8	3.5±2.1
Energy value, kcal/100g	501.4	364.7

As you can see, the energy value of the drone homogenate is even higher than that of RJ. This is due to the higher content of fats and sugars in the homogenate, while the protein content is almost the same.

The molecular composition and structure of these proteins has not yet been studied but, without a doubt, it is rich. Indirect evidence of this is the fact that more than 100 different proteins were found in 2–3 day old drone eggs (Li et al., 2011). At the end of this developmental period, the authors observed the highest protein content with valuable antioxidant properties.

Proteins and amino acids, although they are the main, they are not the only biologically active components in the homogenate of drone brood. To them it is necessary to add uridine, adenosine, guanosine (they are all part of nucleic acids), phosphoric acid and its compounds, glycerol, glucose, inositols (substances with the properties of vitamins), sterols, and fatty acids and their glycerides, as well as sugars, the main of which, as in the case of RJ, was trehalose. This list should also be supplemented with data on the content of vitamins and mineral compounds in the homogenate of drone larvae. Since we did not determine these components, I will give here the data of the Russian company "Tentorium" on the concentrations of some trace elements in the homogenate (mg%): Zn - 5.54, Mn - 4.40, Fe - 3.23, Mg - 2.00. It also contains copper, cobalt, and nickel. These elements, of course, are not contained in the larvae in a free state but are part of important biological catalysts, enzymes.

The homogenate is also exceptionally rich in phosphorus (189 mg%), which is necessary for the synthesis of many biomolecules, such as DNA, RNA, ATP, and phospholipids of cell membranes, among others. The vitamin complex of this product, in addition to the inositols discovered by us, according to the same company "Tentorium", includes (IU/g): vitamin D - 146, vitamin A - 0.083, as well as its precursors (β-carotene and xanthophyll). Vitamins B2 (riboflavin), B3 (nicotinic acid), and B4 (choline) are contained in it in amounts of 0.114, 2.43, and 68.12 mg%, respectively. As you can see, it is especially rich in choline, which in the human body has an important effect on carbohydrate metabolism, regulating the level of insulin in the blood. In addition, choline is a strong hepatoprotector and it is this property of the drone homogenate that was noted by scientists both in Romania and in Russia (Vasilenko et al., 2002; 2005). In Romania, for example, more than 20 years ago the domestic drug Hepatoapimel was recommended for the treatment of liver disease. It has also been shown that extracts from drone larvae have a protective effect on the fetus during pregnancy (Kabała-Dzik, 2007a; 2007b).

Judging by the composition and testimonies of authoritative scientists and apitherapists (Stoiko, 2007), the homogenate of drone larvae is a very valuable beekeeping "product". I have to use quotation marks here because only in very few countries (at least in Europe) are drone larvae or their homogenate actually a product. In others, beekeepers often do not know what to do with the drone brood and how to get rid of it. In large beekeeping farms, this can turn into a problem; many tens or even hundreds of kilograms of this waste must be put somewhere. Of course, that is if it is treated as waste and not as a food product or a raw material for the production of medicines or cosmetics.

The use of larvae, or even whole insects for food is not in the tradition of Europeans, who almost universally suffer from entomophobia. Conversely, in many peoples of Asia and Africa, they are often one of the main sources of an easily assimilated complete protein and have also served as a medicine since ancient times. For example, larvae of the paper wasp (*Polistes dominula*, Vespidae), silkworm butterfly (*Bombyx mori*, Bombycidae), scarab beetle (*Scarabaeus* spp., Scarabaeidae), and Chinese moth (*Hepialus oblifurcus*, Hepialidae) are widely used in Chinese and Korean traditional medicine (Pemberton, 1999). A particularly valuable trophy for honey hunters of wild bees in Asia and Africa has always been the bee brood since it is considered, not only tasty, but also beneficial to health (Krell, 1996).

Drone brood larvae are especially often used in traditional medicine for the treatment of male infertility and menopause in women caused by dysfunction of the endocrine system organs (Meda et al, 2004). A number of studies carried out in recent decades have confirmed the medicinal properties of the bee brood (Iliescu, 1993; Vasilenko et al., 2002; 2005; Gorpinchenko et al., 2004; Seres et al., 2013, 2014). In some European countries, this served as a strong incentive for the start of commercial use, not only as a dietary supplement (Apilarnil Forte and Apivitas), but also as a medicine (Apilarnil Potent® and S.C. Biofarm S.A. in Romania; Ukrainian drug Apidron). In Russia, steps have also been taken to standardize this unconventional beekeeping product (Lazaryan et al., 2003; Budnikova, 2011, Krasovskaya, 2011).

Previously, it was erroneously assumed that the use of drone larvae for the treatment of infertility was associated with their high protein content. In fact, the therapeutic effect is associated with the high content of sex hormones, androgens (Greek ανδρεία – courage and γένο – genus) produced by the sex glands in men, as well as (albeit in smaller quantities) by the ovaries in women and the adrenal cortex in those and others.

According to Burmistrova (1999), fresh drone homogenate contains (in nmol/100 g) 0.31 ± 0.02 testosterone, 51.3 ± 8.7 progesterone, 410.0 ± 65.4 prolactin, and 677.6 ± 170 3-estradiols. These data seem surprising since they showed a higher content of female sex hormones in the male larvae. In a relatively recent study (Budnikova, 2009), the dynamics of sex hormones during the development of the drone from the larval stage to the pupa was demonstrated. Five-day-old larvae contained 8.2 nM testosterone and 2745.0 nM estradiol, however, the ratio of male to female hormone was completely different at further stages of drone development;

a homogenate prepared from pupae 15–17 days old contained 15, 6 nM testosterone, and 343.5 nM estradiol. Nevertheless, it was found that the homogenate of drone larvae had pronounced gonadotropic activity (Burmistrova, 1999). A positive effect on the content of male sex hormones in the blood of the drug Apidron, containing a lyophilized homogenate of drone larvae, was noted in the clinical treatment of male infertility (Gorpinchenko et al., 2004). Similar effects were observed in *in vivo* tests on laboratory animals. Surprisingly, feeding a crude homogenate to castrated rats led to an increase in their weight, not only in the weight of androgen-sensitive organs, but also in the level of testosterone in the blood plasma (Seres et al., 2014).

Another beekeeping waste can also be used for food, the larvae of the large wax moth (*Galleria mellonella*). At least in Asian countries, they are considered edible and moreover, a delicacy. I am far from hoping to inspire Europeans to consciously consume this kind of natural product, although you and I, without noticing it, eat a large number of insects and their larvae, and if we were more tolerant of this, the benefits would be enormous. In support of the benefits of "insect eating", let me cite a statement by the professor of Ohio State University (USA) William F. Lyon:

> *"If Americans could tolerate more insects (bugs) in what they eat, farmers could significantly reduce the amount of pesticides applied each year. It is better to eat more insects and less pesticide residue. If the U.S. Food and Drug Administration would relax the limit for insects and their parts (double the allowance) in food crops, U.S. farmers could significantly apply less pesticide each year. Fifty year's ago, it was common for an apple to have worms inside, bean pods with beetle bites and cabbage with worm eaten leaves. Most Americans don't realize that they are probably already eating a pound or two of insects each year. One cannot see them, since they have been ground up into tiny pieces in such items as strawberry jams, peanut butter, spaghetti sauce, applesauce, frozen chopped broccoli, etc. Actually, these insect parts make some food products more nutritious. An issue of the Food Insects Newsletter reports that 80 percent of the world's population eats insects intentionally and 100 percent eat them unintentionally."*

If we are talking about the drone brood, then there is nothing to disdain at all. Let us recall in what conditions, which are practically ideal from the point of view of hygiene, it is removed, and what it is fed with. What food could be cleaner and more complete in comparison with RJ and a mixture of honey and pollen (or bee bread) prepared by nurse bees?

I have never heard of whether any Western European or North American beekeeper uses drone brood or drone homogenate, at least for personal purposes, but I can testify that many Russian and Ukrainian beekeepers are beginning to appreciate it. They not only eat it themselves and their family members, but women make their own cosmetic masks with it this product. For some, the business went so successfully that from the formula "rational use for destroyed drone brood" to "growing drone brood as a profitable object of a beekeeping business."

Such a large company, as already mentioned, "Tentorium" releases its product in compliance with the technology developed for it. What do other beekeepers, connoisseurs of the homogenate, do? Firstly, many of them prefer to use colonies with two-year-old queens for raising drone broods. Secondly, they use collapsible frames designed by the Ukrainian beekeeper V. Dombrovsky, specially designed for these growing drone broods (Fig. 4.1). These frames can be used in Dadan's hives.

Figure 4.1. Dombrowski frame for hatching drone larvae (left) and frame with sealed brood of drone larvae (right)

Brood frames are usually taken just before the cells are sealed. The larvae, removed by hand, or in a honey extractor, are converted into a homogenate by pressing, grinding, or squeezing and here the most subtle aspect appears, preservation of the homogenate.

Unlike RJ, drone homogenate belongs to perishable products since it does not contain antimicrobial components in noticeable quantities. Therefore, it is immediately frozen and stored at a temperature of about -18°C or thoroughly mixed in different proportions with honey. For long-term (about six months) storage at room temperature, mixtures with a homogenate content of not more than 1–2% are suitable; for storage in the cold, but not frozen, are up to 10%. However, with such a high homogenate content, the shelf life is significantly reduced. For immediate use, even mixtures of the homogenate with honey in a 1:1 ratio can be used. Preserving with honey is the easiest and most reliable way to preserve the biological activity of this product. Of course, the dried, lyophilized homogenate can be stored the longest. This powder is used for the production of BILAR series preparations developed in the Ukraine. L. A. Burmistrova (1999) proposed that for long-term storage of the homogenate to bind it with an adsorbent, consisting of glucose and lactose, similar to what is done in the case of RJ.

In conclusion, a few words about one more possibility of using the drone homogenate. In the late 19th century, the famous American beekeeper, Amos Ives Root (1839–1923), recommended squeezing drone combs with a press, preserving them with sugar syrup, and feeding the bees to stimulate family development, especially in the spring. This feeding has been tested by some Russian beekeepers and gave good results. Some of them use a mixture of homogenate with sugar syrup in a ratio of 1:10 or an even higher homogenate content. Others introduce pure homogenate into the feeders without adding syrup. In this case, it is necessary to strictly observe the requirements of hygiene and prevent spoilage of the feed. The homogenate stocked up from last year and frozen in the spring can also be used for the preparation of pasty protein dressings. Notably, feeding bees with a drone homogenate will not harm them in any way. It is known that under certain conditions, with long-term bad weather and the associated lack of bribe, it becomes necessary to drastically reduce the number of broods in the family. This is achieved as a result of cannibalism, bees simply eat the youngest larvae.

Chapter 5 Propolis: *urbi et orbi*

5.1. Propolis is more than glue

Human acquaintance with propolis goes back centuries and now one can only guess when and how the medicinal properties of this product were discovered. Thanks to these properties, it has been highly valued by many peoples belonging to different cultures. It can be assumed that a person guessed about the protective properties of propolis, discovering it in bee nests along with the "incorruptible relics" of uninvited lovers of honey, such as mice and lizards, that were killed by bee venom. Bees cannot take these remains out of the nest and they cannot be left to remain as they will rot and poison the nest. This is where propolis comes to the rescue; the dead bodies of the "invaders" covered with a layer of this substance are not subject to decomposition, which is the best evidence of its surprising antiseptic effect.

Collectors of honey from "wild" bees living in hollows of trees or in crevices of rocks also saw that the honeycomb was firmly glued to the walls by the same material.

It is also easy to see that the inner surface of the hollow is covered with a thin layer of propolis, which prevents further destruction of wood by ubiquitous microscopic fungi and also prevents water and tree sap from seeping into the nest. This wood moisture could have a bad effect on the "microclimate" in the nest and getting rid of it would require a lot of effort on the part of the toilers, who already have other duties. Consequently, propolis also serves as a kind of internal building material for bees.

The characteristics of this sticky substance, understanding of its nature, and the possibilities of its use in different cultures were different. Its use for medicinal purposes are known because of the works of outstanding scientists of ancient Greece and Rome, such as Aristotle and Pliny, also known as the Elder. It is likely that Aristotle, who at one time was engaged in "medical practice", used propolis as a medicine or as an integral part of it.

Cultural beekeeping originated in southern Europe, on the island of Crete, in Greece, and Rome, as well as in northern Africa and Asia Minor. People kept bees in hives of various types: reed, straw, or clay. This type of farming made it possible not only to collect more bee products, including propolis, but also to better observe the life of bees than is possible when extracting honey from wild bees from natural or artificially created hollows. The person engaged in this usually limited themselves to collecting honey and did this by cutting the honeycomb located in the hollow of the tree with a knife and at the same time, they could generally avoid contact with propolis. The beekeeper was typically content with honey and wax and not very interested in propolis. Naturally, the level of knowledge about the origin and role of propolis in the life of bees was different among the "cultured" beekeepers of antiquity and among people who were engaged only in collecting bee products.

Therefore, it is not accidental that this substance was named differently in different parts of Europe. The Greek word "propolis" in translation means "the threshold of the city" and in this name you can easily see a hint of its defensive function. Moreover, in ancient times, cities were surrounded not only by walls but also by other engineering structures to protect against invasion, such as ditches filled with water and suspension bridges, among others.

Here is an example of the use of propolis for «defensive purposes» by a dwarf bee, Apis florea. This bee nests in the open air, attaching its small honeycomb to the branch of a tree or shrub with wax and collecting propolis in small quantities for other purposes. Here is how Professor E. Wilde describes in the book «Beekeeping» (1998) the use of propolis by this bee for defence: «The open position of the nests of the dwarf bee in the bushes, near the surface of the earth, has developed in her a kind of defensive strategy against small robbers, in particular against ants ... A very effective, sophisticated remedy is the "sticky traps" located on the branch on both sides of the nest. They consist of propolis, which, thanks to the care of bees, remains soft and sticky all the time. This substance repels ants, but at the same time can stick most cheeky robbers.»

The northern peoples, including Slavs, Germans, and Anglo-Saxons, began to breed bees much later and in their view, the substance in question plays mainly, if not exclusively, the role of internal building material, such as glue and putty. Hence, has been called kit pszczeli (bee putty) by the Poles, bee glue by the British, and kittharz (gum putty) by the Germans. Now these names are considered outdated and the widespread use of the word "propolis" is evidence of its recognition (although, perhaps, not fully realized) and more complex functions than only construction and repair, such as protection.

So, what is the protective power of propolis and what does it give the bees themselves? These are the main questions that I will try to answer, relying primarily on the achievements of many, many scientists, as well as on the results of my own research. First, however, it is worth paying attention to where and how the bees collect the material for the manufacture of propolis.

5.2. The work of bees to collect herbal balms: Workaholics and lazy people

Everyone knows that honey and beebread do not "grow on trees" but are produced by bees. Honey from the sweet nectar of plants or honeydew and pollen is used as plant raw material for making beebread. As for propolis, even in scientific papers, you can often read that it is just plant resin or a balm brought by bees to the hives. There are also works that state that propolis is a product of the biosynthesis of bee glands and, thus, animal origin is attributed to it. Moreover, there is still no complete clarity on this issue.

According to Mark Twain, "the information that the ancients did not have is very extensive," nevertheless, they were able to observe and comprehend what they saw. Starting with the most ancient hypotheses, the aforementioned Gaius Pliny the Elder, in his monumental work "Naturalis Historiae", wrote that the precursor of propolis is the gum secretions of many plants collected by bees, among which he mentioned poplar, willow, and grapes. Judging by some publications, earlier Aristotle spoke about the "external" source of propolis. It is unknown what was judged on this score in the Middle Ages but it seems that no one put forward other hypotheses. At the beginning of the 19th century, the Swiss naturalist François Huber (1814) described an experiment he invented to observe the collection of resinous secretions by bees on the surface of poplar buds; in modern literature, these secretions are called *exudates*, and in English literature, the word *resin* is most often used. In the immediate vicinity of the hive, a "bouquet" of cut-off poplar branches with large buds richly covered with aromatic resin was placed. The bees reacted with great attention to this "gift" and Huber and his assistant, according to his testimony, had the opportunity to observe how they collect exudates[11].

This operation was described in more detail 150 years later by a number of other researchers (Haydak, 1953; Meyer, 1956). Observations have shown that bees are very picky in their choice of buds; they fly from one branch of a tree to another, alternately touching the tops of the buds with their antennae, as if checking the quality of the material. It would seem that there is no difference as all the buds are covered with a sticky mass. It would seem that it would be more efficient to collect it without wasting time and carry it to the hive. This behavior, obviously, has a deep meaning as the chemical composition of bud exudate, even on the same tree, can be different. For example, due to the volatilization of some of the compounds on the sunny side of the crown or even due to undesirable (from the point of view of collectors) chemical processes under the influence of the same sunlight.

When the choice is made, the collection operation begins directly, consisting of several stages. First, the bee tears off a piece of resin with the help of its upper jaws (mandibles) and chews it, adding saliva to form a small lump. Then, with its front legs, it takes the lump from the mandibles and transfers it to one of the legs of the middle pair. Finally, the resin is transferred to the basket of one of the legs

---

[11] *François Hubert (1750-1831) became blind at a young age and all the brilliant experiments he invented were carried out and described with the help of his assistant, François Burnens.*

of the third pair, on the same side of the body. These steps are repeated until both baskets are full. After that, the bees take off but not all of them immediately go home with their bounty. Some rise into the air for only a few seconds and again return to their favourite source to replenish their baskets with fragrant resin. It is likely that during this short flight, the bee checks the weight of the collected load and if it seems insufficient, replenishes it.

These observations from the time of Aristotle to the second half of the 20th century were made "with the naked eye." In recent decades, more diverse and sophisticated techniques have come to the aid of researchers. A microchip fixed on the back of the bee makes it possible to follow it throughout its entire life span and thermal imagers and video cameras allow the observation of life inside the nest even in complete darkness, localizing the place and type of work carried out in different parts of the hive, and registering changes in body temperature of a single bee when performing various operations, among other observations.

The use of some of the latest observation techniques by researchers at a field station at Cornell University (USA) helped to study in detail the actions of bees involved in both the delivery and further use of balsamic resin. In the course of video observation of the newly born bees and provision with "personal" tags*[12], it was found that some of the resin foragers brought their load not only in baskets but also in mandibles. Of the 35 monitored resin foragers, this super-enthusiasm was noted in five individuals (more than 14%) (Nakamura & Seeley, 2006). However, at the same time, among the resin foragers, sloth bees were also found, bringing home incomplete baskets. In a fairly representative group of 180 resin collectors, a comparison was made between the diameter of the load carried in the basket (D) and the average diameter of the basket itself (W) and according to this criterion, three groups of collectors were identified. Incomplete baskets (D < 1W) were brought by 16% of resin foragers; in 62% of bees, the diameter of the load was within 1–2 W, and in 23% of the female workers, D > 2W. In a word, an almost normal distribution described by a Gaussian distribution curve of random variables.

---

12 *The French natural scientist René Antoine de Réaumur (1683–1757) was the first to mark bees with paint for further observation. He was probably the first scientist to use the technique of active experimentation in the study of bee life. In particular, he found out many details of brood development and care; the temperature in the hive was measured using an alcohol thermometer, which he invented. He published the results of his observations and experiments in 1740.*

The time to it takes to fill both baskets (each of them weighs about 10 mg on average) can be very different, from 6–7 minutes to an hour or even more. It depends on various factors, such as the state of the weather (strong wind or relatively low air temperature does not favour collection) or the wealth of the source.

Observations show that bees are most willing to procure raw materials for propolis by the collection of plant exudates on sunny days starting at around 10 am for about 14–15 hours. In the middle latitudes of the Northern Hemisphere, most often this type of work occurs in the second half of the summer and early autumn, although in some cases it is observed even in May. Many researchers are now inclined to believe that there is a seasonal nature to propolis harvesting and that it shifts between July and September. This may be explained by the completion of the main honey harvest when the influx of nectar and pollen into the hives decreases. During the main honey harvest period, according to this point of view, bees have a higher degree of motivation to collect these particular gifts of nature. However, such judgments should be treated with a certain degree of scepticism, if only for the reason that only a small part of the workers are employed in the collection of resins and balms and the distribution of material inside the hive; their distraction from flowers does not affect the situation in the family. Most likely, the activity of bees in collecting resins is determined by nothing more than the current need of the family for propolis.

If the collection of raw materials necessary for the preparation of propolis is described in sufficient detail, then its further placement and use in the hive has been studied to a lesser extent. Even less is known about ways to inform family members about propolis needs. Research at Cornell University sheds some more light on these issues. First of all, earlier reports (Rösch, 1927; Meyer, 1956) confirmed that the unloading of baskets is carried out by the receiving bees not at the entrance to the nest (as is the case when taking nectar and water), but near the places of its subsequent use or "warehousing".

In the process of unloading the baskets of one collector, two to six receptionists (worker bees) take part and sometimes two or even three bees partake in this simultaneously. Interestingly, according to the observations of the authors Nakamura and Seeley (2006), the duration of unloading the baskets was inversely proportional to the size of the load brought. At $D < 1W$, it took on average 55 minutes; at $1W \leq D \leq 2W$, it decreased to 47 minutes; and at $D > 2W$, up to 40 minutes. It seems that the receptionists are work faster to unburden the more hardworking sisters, whom they relieve in the first place.

Non-flying, hive bees, which we will call here "plasterers", are engaged in unloading the baskets. As already mentioned, this happens in the immediate vicinity of the place of use or storage, which is very wise as it saves time and effort. They pinch off a piece of resinous substance, chew it thoroughly, mixing it with saliva, and immediately attach it to the wall of the hive or to the frame. It is interesting that the bees who brought their prey in mandibles immediately go to the place of use and without waiting for the baskets to be unloaded, use it for their intended purpose, thus performing the work of "plasterers". Other resin foragers, freed from resin, were also noticed doing the same job. This plant material, twice chewed and soaked in saliva, is called propolis (Simone-Finstrom & Spivak, 2010).

There is no direct evidence that anything else has been added to the resin at this stage but it cannot be ruled out that it is then that wax also gets mixed into it. Earlier it was believed that old bees with already atrophied wax glands were engaged in the "plastering" work (Meyer, 1956). However, Nakamura and Seeley (2006) refuted this opinion in a controlled experiment; according to their observations, "plasterers" were younger than resin foragers and began work in this field at the age of 14 days, moving into the category of resin foragers only on the 25th day of life. This means that "plasterers" are, in principle, capable of producing wax and adding it to plant materials supplied by older workers.

The described experiment also revealed a number of significant differences in the collection of nectar and resin. First, it turned out that there was no strict division in the distribution of responsibilities between the bees delivering the resinous material and its acceptors-plasterers; the same bees were seen in both types of work within the same day. Secondly, recruiting dances are performed by nectar resin foragers on combs near the entrance, among the unemployed bees, who huddle there like at a labour exchange and resin foragers dance deep in the nest, among the sisters who are "experienced" in the use of propolis. Third, resin collectors are not exceptionally faithful to this occupation. After one or several days of work in this field, many of them switched to collecting pollen or nectar and some then returned to their original occupation.

There is one more important issue concerning "information support" of the collection of resinous material. It is unknown how resin foragers are informed about the need to continue or stop this work. The authors of the described experiment (Nakamura & Seeley, 2006) did not come to a final conclusion on this score and considered two equal hypotheses. The first of them, "unloading" presupposes an indirect mechanism for obtaining information; the slower the unloading of resin from collectors' baskets, the less need of the family for this material. According to Meier (1956), emptying the baskets can take as long as two days. It is clear that such a delay would discourage any desire to continue collecting balms. Incidentally, it is this indirect mechanism that informs female workers delivering nectar and water (Kühnholz & Seeley, 1997).

The second hypothesis, which can be called "sensory", is based on the fact that many resin foragers, after unloading, were noticed in active inspection of places for the potential use of propolis; with their antennas, they touched cracks and various kinds of irregularities, examining them in this way, or even participating in "plastering" works. In this case, they get the information they need directly. One way or another, one cannot disagree with the following conclusion: "Both (hypotheses) may be correct, and if this judgment is correct, it follows from it that social insect workers, despite their small brains, are able to accumulate and summarize information from different sources to increase their awareness of the situation in their family"(Nakamura & Seeley, 2006).

The ability to accumulate information, that is, to store it in memory, and, moreover, to generalize and transmit it to other individuals of their own species using various means of communication it's a sign of intelligence, isn't it?

Every study and experiment has a specific purpose, boundaries, and limitations. As part of the study described above, purposeful observations were made from the delivery of balms and use of propolis by bees to seal cracks. By removing the propolis accumulated in the hive daily, the authors artificially forced the bees to continue this activity. Naturally, other ways of using propolis fell out of their field of vision, such as covering the inner walls of the cells with it, in the preparation for laying eggs, or encrusting the edges of wax cells with propolis to give them greater strength (F. Huber wrote about this use of propolis by bees as early as 1814). Furthermore, these observations do not tell us anything about, perhaps, the most important function of propolis, which are its protective functions. This is entirely determined by the chemical composition of propolis, and hence, the composition of its plant precursors. Before moving on to these questions, there is still the problem of the origin of propolis.

The ancient Greeks and Romans believed that propolis was a material of "external" origin. However, this idea was revised at the beginning of the 20th century by the German researcher and beekeeper M. Küstenmacher (Küstenmacher, 1911).

According to him, the reason for this was that "... despite careful observation, ... not a single bee could be noticed that would collect resin from the buds of plants." It is likely that the observations of F. Huber escaped his attention, who succeeded nearly a hundred years ago prior to this time. As an argument against the "external" source of propolis, Küstenmacher also cited the following reasoning: propolis is found in hives not only in spring and autumn but also in summer; in the summer, the buds have already blossomed and, therefore, there is nowhere to take this very resin.

Without putting it in a distant box, I will immediately note that the argument about the absence of a source of resin in early summer does not correspond to reality. Leaf buds are laid very early by many tree species and in many species, they are richly covered with exudates from the very beginning. I have observed, both in Poland and north-east Latvia, exudates on poplar buds even at the end of May.

According to Küstenmacher, propolis is produced in the stomach of the bee, in its front section, called the honey sac, and its plant precursor is plant pollen. Further "development" of these concepts was the hypothesis of the dual nature of propolis, for two different types of this product, called "false propolis" and "true propolis" (Ioirish, 1976). According to a new hypothesis, "false propolis" is the notorious bee glue, collected from the buds of trees and mixed with wax. Moreover, it performs repair and construction functions, serving for filling gaps and attaching honeycombs, among other functions. The "true" propolis, as suggested by Küstenmacher, is supposedly formed from pollen and its "... bees belch themselves every time they eat pollen. The shell of pollen grains contains a certain amount of resinous substances and balsams released during digestion from pollen grains" (Ioirish, 1976). According to the supporters of this hypothesis, it is this "true propolis" that is used by bees to disinfect the nest and insides of the cells prepared for raising offspring. They even gave this "oily, balsamic liquid" a special name: "pollen balsam" (Kędzia & Hołderna-Kędzia, 2009).

The final confirmation or refutation of certain ideas about the nature and origin of propolis should be sought by comparing the chemical composition of both the alleged plant precursors and propolis itself. As for the Küstenmacher hypothesis and its further modifications, an attempt to precisely test this was undertaken back in the late 1970s by S.A. Popravko (Russia). It consisted of studying the composition of pollen and propolis collected by control colonies of bees. As a result of repeated experiments, they failed to find the slightest similarity in the composition of the flavonoids of both of these products. Hence the conclusion of the authors, "Thus, the negative result of the experiment showed the *groundlessness of the assumptions about the origin of propolis from pollen grains*" (italics, Popravko, 1982).

Popravko and his collaborators investigated the composition of the notorious "imaginary propolis" that they collected from the framework and the conclusion about the absence of its connection with pollen is undeniable. However, perhaps this propolis is too rough for the delicate skin of the larvae and, therefore, "pollen balm" is really used to prepare the insides of the cells for brood by bees. Finally, it would be necessary to try to somehow obtain washings from the walls of the cells and, on the basis of their chemical analysis, establish the nature of "true propolis".

Moreover, knowledge of the chemical composition of propolis is necessary to understand its protective properties, as well as the possibilities and limitations when used for medicinal purposes.

5.3. Is it possible to establish the origin of propolis?

The chemical composition of propolis, as its researchers have long been convinced, is unstable. Even when collected in the same apiary, and even from adjacent hives, it is never completely identical in composition (Greenaway et al., 1988). This circumstance is surprising in itself and no explanations have been found for it until now. The differences in the composition of propolis from different regions and different countries are all the more striking and this can be seen by comparing the "profile" of the chromatograms shown in Fig. 5.1.

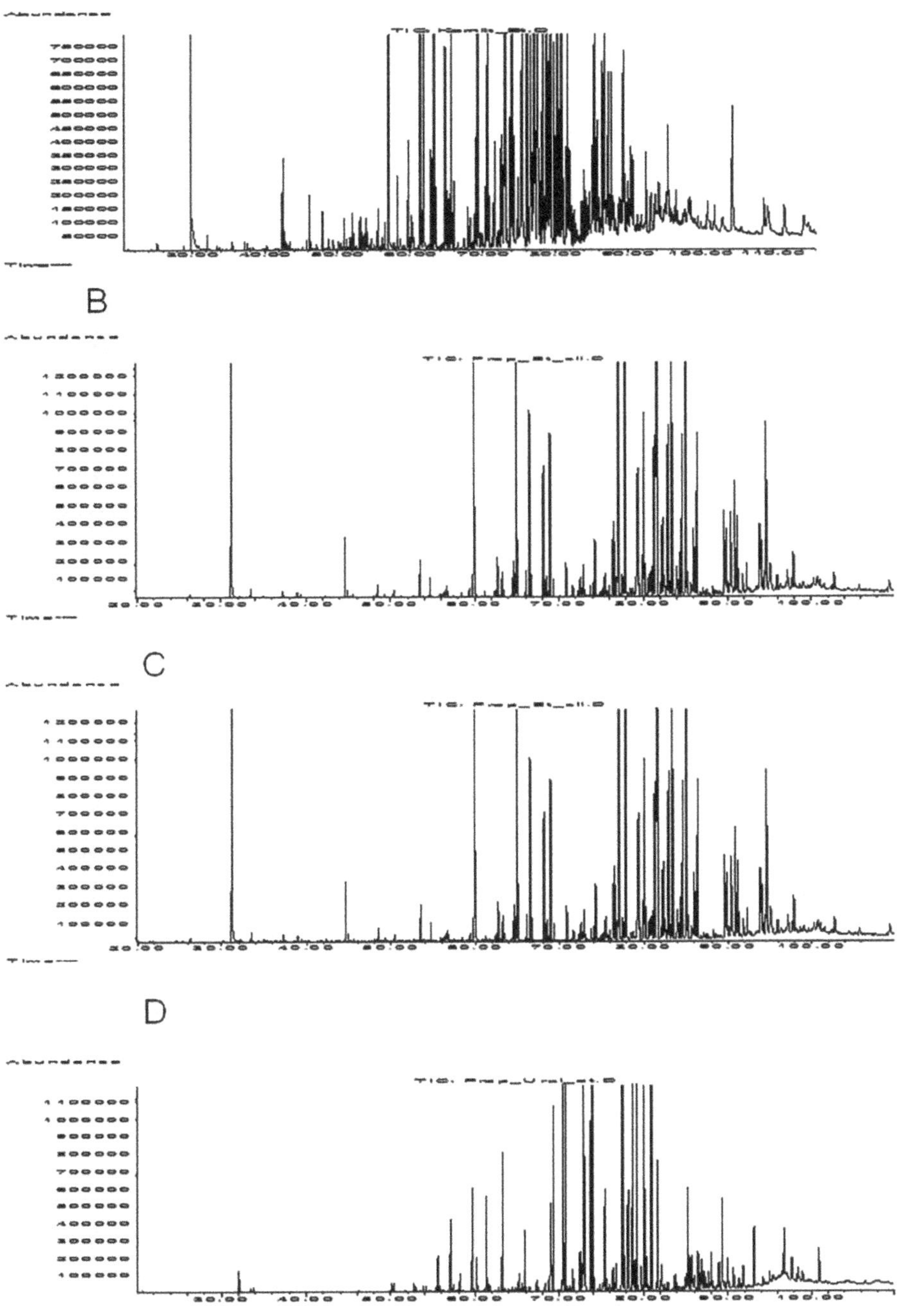

Figure 5.1. Chromatograms of ether extracts from propolis collected in Poland (A - Kurnik, Wielkopolskie Voivodeship; Socha beekeeping farm); B - Struzhe, Malopolskie voivodeship, beekeeping farm "Sędecki Bartnik®"), in the north-east of Latvia, Tirza (C), and in Russia, Yekaterinburg region (D).

In addition, considerable discrepancy in reports on the composition of propolis is associated with differences in the approach and equipment used in its analysis by different authors. There is no such technique or analytical approach that would make it possible to determine at once the entire range of substances forming propolis, from the most volatile to very low volatile, from non-polar to highly polar. Some of them dissolve well in non-polar hydrocarbons, others in slightly more polar solvents, such as diethyl ether and ethyl acetate, and still others only in strongly polar water-alcohol mixtures. Usually, several groups are distinguished as part of propolis (Table 5.1), each of which, in turn, consists of a variety of compounds that are diverse in their chemical structure.

Table 5.1. Propolis constituents and their relative (%) content

Constituents	%
balms	50–80
essential oils	4–15
waxes	12–40
pollen	5–11
plant pigments	4–10
mechanical impurities	5–20

As you can see, the relative content of individual groups fluctuates within fairly wide limits and depends on various factors. It is believed, for example, that bees add more wax when, for one reason or another, there are few plant balms in the environment, or when it is difficult to collect them. This may be due to hot weather, when the balms dry out, or cold weather, when they are almost hard. With the exception of mechanical impurities, whose content to a large extent depends on the diligence with which the beekeeper wields a knife to scrape propolis from the frame, each of the above groups consists of a great variety of compounds, of which only a relatively small part has been identified so far. Review works usually speak of the presence of 150–350 compounds in propolis but it certainly would not be an exaggeration if we say that there are more than a thousand of them, even if we limit ourselves only to "European" propolis. For those readers interested in the composition of propolis from other continents, we refer to earlier (Marcucci, 1995) and relatively recently published reviews (Toreti et al., 2013; Bankova et al., 2018; de Pontes et al., 2018; El-Guendouz et al., 2019).

Proceeding from the assumption that the main precursor of propolis is balsamic exudates from plants, we endeavour to identify any "indicator" compounds for exudates of certain plant species, that is, those that would unambiguously point to a specific species.

In the literature, various types of poplar and birch are called "propolis-giving" plants of the temperate zone of the Eurasian continent:

- black poplar, *Populus nigra*,
- Italian poplar, *P. nigra* var. *piramidalis*,
- various hybrids of *P. nigra* and American poplar species,
- aspen, *Populus tremula*,
- silver birch, *Betula pendula* (its outdated name is *B. verrucosa*),
- downy birch, *B. pubescens*,
- black alder (*Alnus glutinosa*),
- horse chestnut (*Aesculus* L.),

as well as beech (*Fagus* spp.), elm (*Ulmus* spp.), pine (*Pinus* spp.), and grapes.

As for poplars, birches, and horse chestnuts, they undoubtedly can serve as a source of balms for bees, since their buds exude abundant exudates, as can be seen in Fig. 5.2.

Figure 5.2. Exudates on the buds of black poplar and horse chestnut. (Photo: V. Vinogorova)

Black alder is not so generous in this regard but its buds and young leaves are sticky (hence the word "glutinosa" in its name). In the case of beech, elm, and pine, I have great doubts about their "propolisity". At none of the stages of bud development in European species of beech (*F. silvatica*) and elm (*U. campestris*), can we see even traces of exudates on the buds. It is not clear from the literature what specific types of beech and elm are in question; perhaps, it is found in the eastern or Balkan varieties of beech (*F. orien*talis) and in the Central Asian species of elm *U. densa*, or cork elm (*U. suberosa*). There are no gum secretions on the buds of the white poplar (*Populus alba*) either, although this species is also mentioned in some publications as a source of bee balm.

The collection of resin by bees from buds or stem secretions, which leaks from wounds on the trunks of coniferous trees, such as different types of pine, is also in question. This source is undoubtedly rich but somehow it is not possible to find the main indicator components of these resins in the "northern" propolis. These are diterpene acids, called resin acids, such as abietic, dehydroabietic, isopimaric, and pimaric acids, among others. Together, they form more than 70% of the pine resin mass. However, out of more than 100 propolis samples (from the Pacific Ocean to the Baltic Sea, from east to west, and from the White to the Black Sea, from north to south), only two have contained trace amounts and only one of them was dehydroabietic acid. One of them was collected on the banks of the Chusovaya River in the Perm region and the other, in the north of the Vologda region, that is, in the middle taiga zone with predominantly coniferous vegetation. Another propolis sample containing small amounts of the same acid originated from the northeast coast of the United States.

Not so long ago, diterpenoids, related to the aforementioned resin acids, were discovered in the composition of propolis from Sicily (Bankova et al., 2002), parts of Greece, and the Mediterranean islands, including Crete, the Aegean Sea Islands, and Cyprus (Melliou & Chinou, 2004; Kalogeropoulos et al., 2009, Popova et al., 2009). In Sicilian propolis, the content of diterpenic acids was 53.2%. The most likely plant precursors of these compounds in most of these southern regions are cypress trees, especially the evergreen cypress (*Cupressus sempervirens*). In Crete, as well as on some large islands of the Adriatic Sea, there are no birches and it is very rare to see poplars outside the cities with their parks and alleys, and there are no other supposed "propolis-giving" deciduous trees. Cypresses of the species *C. sempervirens* (as well as Mediterranean pine species, *Pinus pinea*, *P. nigra*, *P. halepensis* and *P. brutia*) are quite common here. Chromatographic analysis of resin collected by me on the Croatian islands of Brač and Zlarin, from cracks in the bark of cypress trees, fully confirms the hypothesis of Greek researchers (Melliou & Chinou, 2004) that the resin of these trees are the main precursor of "Greek" propolis. Notably, in propolis from the coast of Croatia, on which, along with poplars, many cypresses also grow, the "cypress signal" completely disappears. This suggests that the resin of conifers only attracts bees if there are no deciduous trees ready to offer them balsamic secretions. On most of the Eurasian continent, there is no shortage of such trees.

Before we start discussing the composition of exudates and propolis, let us make a short "excursion" to the field of analytical chemistry of complex natural objects. The most effective and, therefore, the main method in such studies is gas chromatography (GC), which is distinguished by the highest (in comparison with other methods) separation power and sensitivity. The "product" of the analysis is a chromatogram, a series of compounds separated in a very long and very thin capillary column (as a rule, the length of such a column is 30 meters, and the inner diameter is only 0.25 mm), presented as peaks of different heights. The higher the peak of a compound on the chromatogram, the greater its content in the analyzed mixture.

In Fig. 5.3, you can see the chromatograms of exudates from the buds of some trees and that their composition is rich. Each of them contains hundreds of components, in various quantities from "trace" (less than 0.05%) to very representative. Trace components of propolis exudates are not unimportant; some of them are "indicators" for a given plant species and the total content of trace components can have a decisive effect on biological activity and, thus, on the medicinal properties of propolis. The important role of "traces" is convincingly evidenced by the experience of American scientists, who, in the 1970s, tried to recreate the aroma of strawberries on the basis of data from the composition of many dozens of the main volatile compounds previously discovered in it. By mixing them in the right proportions, the researchers got something with a "scent" ... of burnt rubber. It turned out that the pleasant smell of strawberries is determined precisely by a set of trace compounds. Therefore, it is important to identify as many components as possible in the objects of our research, even if some of them refer to minor components.

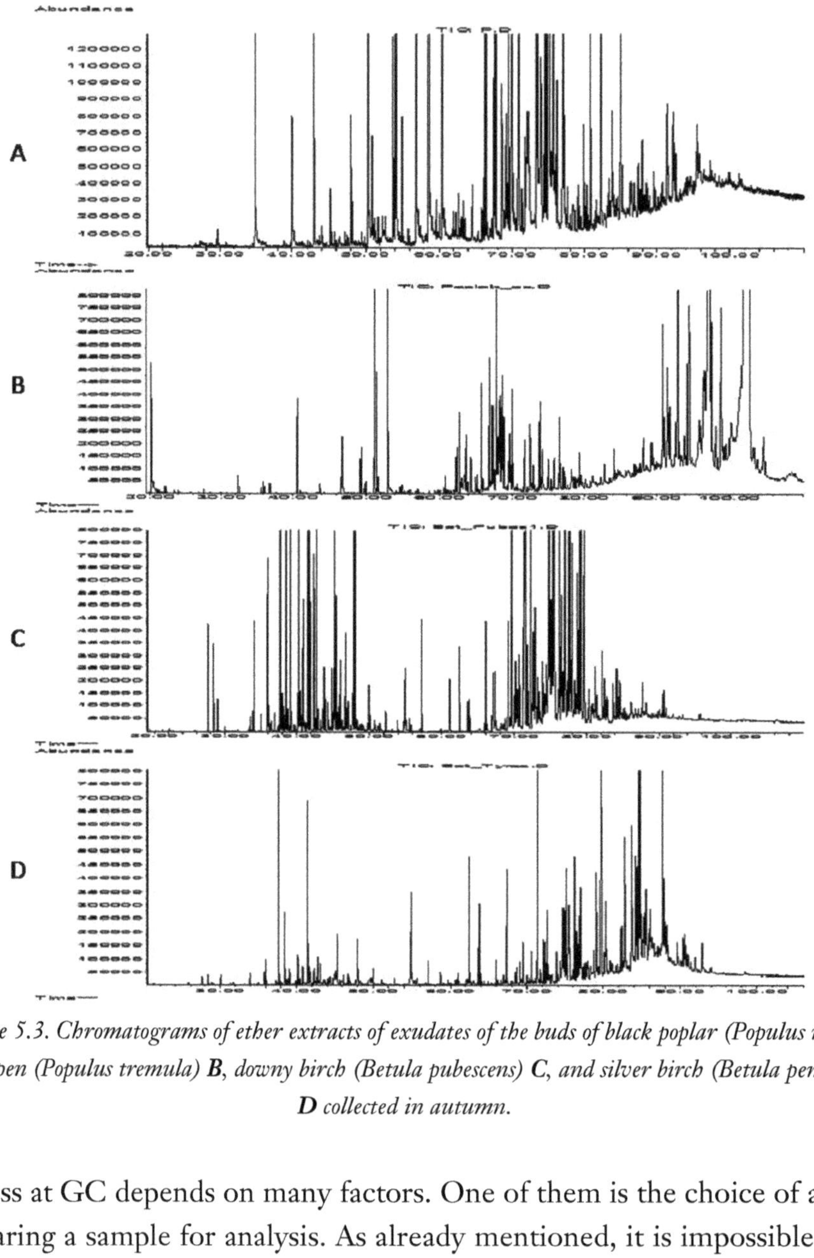

Figure 5.3. Chromatograms of ether extracts of exudates of the buds of black poplar (Populus nigra) ***A****, aspen (Populus tremula)* ***B****, downy birch (Betula pubescens)* ***C****, and silver birch (Betula pendula)* ***D*** *collected in autumn.*

Success at GC depends on many factors. One of them is the choice of a method for preparing a sample for analysis. As already mentioned, it is impossible to cover the entire range of compounds that form the objects of interest to us in one cycle of analysis. Therefore, the analysis procedure begins with extraction with a gas or a suitable solvent, such as ether. The obtained extract, after its concentration (evaporation of most of solvent), can be introduced into a gas chromatograph and the "spectrum" of the separated compounds can be recorded. However, such an extract contains substances of different polarities, and the most polar ones turn out to be "invisible" (as if covered by an invisible cap), since they cannot pass through the column and reach the registering device (detector).

Fortunately, chemists have learned to bypass this obstacle through derivatization, that is, the conversion of polar and non-volatile compounds into less polar and volatile forms. In our case, the most convenient way of derivatization (for example, some compounds of ROH) was silanization, according to the following simplified scheme:

$$RO\text{-}H + (CH_3)_3Si\text{-}OR' \rightarrow RO\text{-}Si(CH_3)_3 + H\text{-}OR'$$

The product of the reaction, trimethylsilyl derivative RO-Si $(CH_3)_3$, is abbreviated as TMS. Chromatograms in Fig. 5.4 show how many "invisibles" were able to remove their "hat of invisibility", including phenolcarboxylic acids and flavonoids, using this approach. However, all components must be identified, as well as their content.

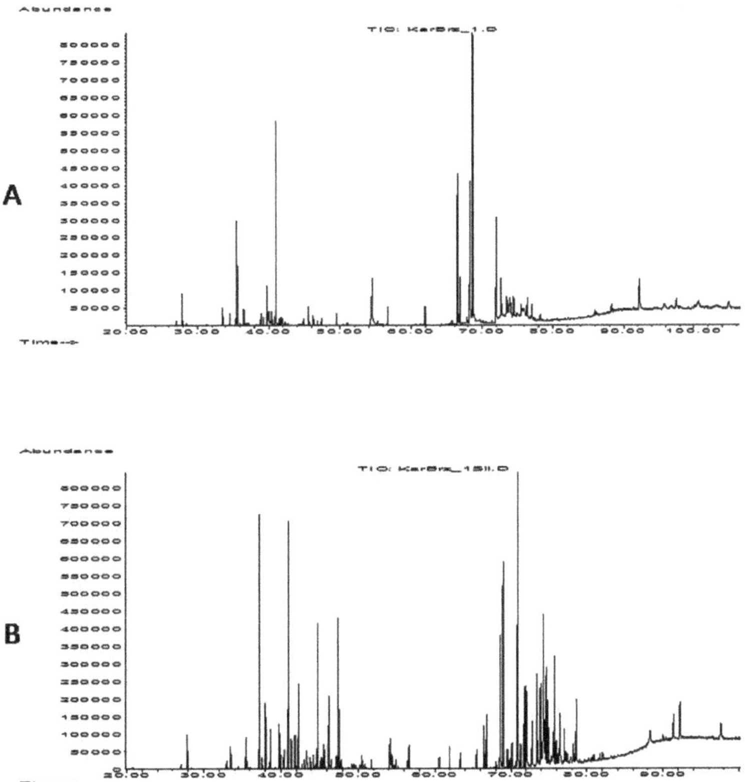

Figure 5.4. Chromatograms of ether extracts of birch winter bud exudates before (A) and after silanization (B).

A combination of gas chromatography and mass spectrometry (GC-MS) has proven to be the most suitable for identification of organic compounds. The gas chromatograph makes it possible to separate compounds and the mass spectrometer (MS), in this "tandem" system, serves as a detector. MS provides the opportunity to identify each of the separated compounds. The identification is based on a comparison of the mass spectrum recorded during the analysis with spectra contained in a computer database. This process can be compared to the work of a border guard, who looks at a person crossing the border, then at his passport photo, making sure who he is dealing with. However, not everything is always so simple.

First, the databases of mass spectra are far from complete and a particular deficit is felt precisely in the case of TMS spectra. In such a situation, the experience and intuition of the researcher helps, but intuition alone will not go far. Therefore, we constantly replenish our own, auxiliary database of mass spectra of compounds identified at different times and in different samples. Secondly, there are often cases when unambiguous identification occurs. Solely by mass spectra, it is generally impossible due to the existence of twin compounds, or even "twins" with indistinguishable "photographs" or mass spectra of a similar chemical composition (for example, isomers and homologues). In cases where these are remarkably similar, it is easy to make a mistake. In such cases, the border guard would need additional data to identify the traveller, for example, the special features of the traveller contained in the computer database and in the passport. In GC-MS analysis, so-called retention indices, RI, characterizing the position of the compound peak on the chromatogram relative to specially selected witness substances, are used as "special features"; specialists call them additional identification parameters. These very important parameters are also collected in a special database (Isidorov, 2020).

All these difficulties have been overcome and identification has been successfully carried out. In the exudates of the buds of three representatives of the genus *Populus* (poplar), 269 compounds were identified and 379 substances in two birch species. There is no need to include all these substances here. To get an answer to our question about the "indicator" compounds, it is sufficient to summarize the data presented below in graphical form.

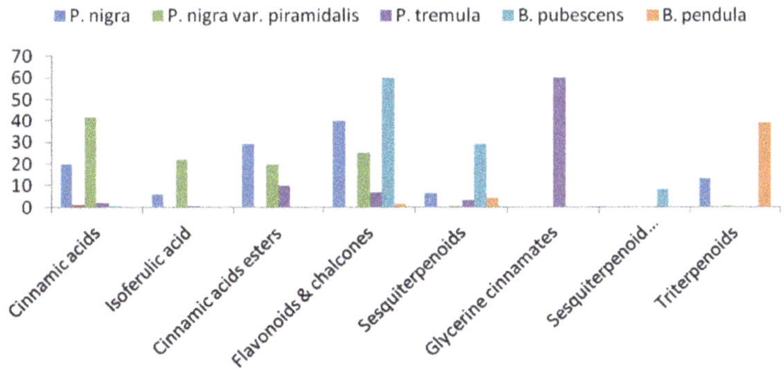

Figure 5.5. Group composition of bud exudates of some plants

The groups of compounds not shown on the graph, such as fatty alcohols and acids and aromatic acids and their esters, are mostly not specific. They are found in exudates of all plant species and their content is low. However, it is easy to find differences in the content of so-called polyphenol compounds, which include various flavonoids and cinnamic acids and their derivatives. It is easy, for example, to distinguish aspen exudates from exudates of two other poplars, downy birch from silver birch: aspen is characterized by a high content of "atypical" cinnamic acid glycerides (glycerine cinnamates). Our studies made it possible to identify as many as 39 compounds of this class in aspen buds (Isidorov et al., 2008). The presence of esters of the same cinnamic acids and sesquiterpenols ($C_{15}H_{26}O$) is a distinctive feature of downy birch exudates. The second species, silver birch, does not contain these compounds at all, but is rich in triterpenoids, primarily dammaradien-3-one (1) and dipterocarpol (2).

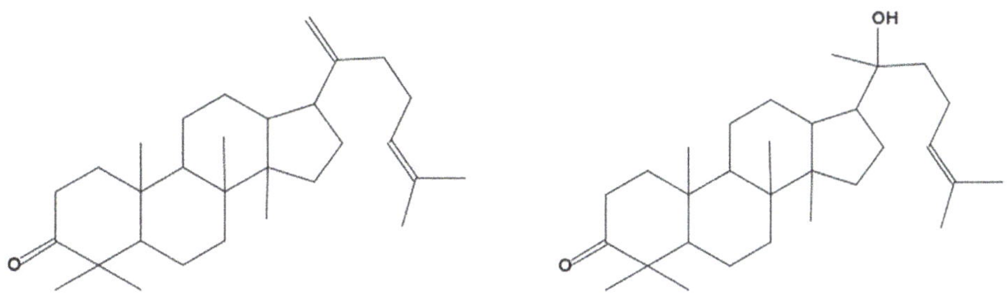

1 2

Figure 5.6 . Chemical formula of 3-dammaradienol (1) and dipterpcarpol (2)

It is more difficult to distinguish between poplars, which is not surprising, since "black" and "Italian" poplar are two forms of the same species (*Populus nigra*). Even in this case, the task is not impossible if we pay attention to the content of cinnamic acids and triterpenoids in their exudates. Thus, the buds of the "Italian" poplar secrete large amounts of isoferulic acid and its esters, which the second species does not have.

We also mentioned horse chestnut and black alder as potentially "propolis-giving" trees. As it turns out, horse chestnut bud exudates contain only small amounts of polyphenolic compounds. As a result of GC-MS analysis, we were able to detect only three flavonoids (epicatechin, catechin, and kaempferol), the total content of which barely reached 2.5%. The presence of all six even numerical 3-hydroxy acids C_{12}–C_{22}, the total content of which was at the level of 13–15%, can be recognized as characteristic of it. The exudates of buds and young leaves of black alder turned out to be even less rich in polyphenols. The greatest amounts in them were the glycerides of fatty acids (palmitic, stearic, oleic, and arachidic), of which, fats, including vegetable fats, are composed. The content of these glycerides in exudates exceeded 60% (Isidorov et al., 2016). It is likely they that form the sticky substance that covers the buds and young leaves of the alder.

Therefore, we have identified indicator compounds and we can now try to find a "trace" of plant precursors in propolis. We would have to search for them among 519 compounds and thus far, we have been able to identify many diverse substances in the analyzed Eurasian propolis samples. Below is shown, in graphical form, only a small part of the results of our research, nevertheless, these data fully reflect the general situation with propolis on this continent.

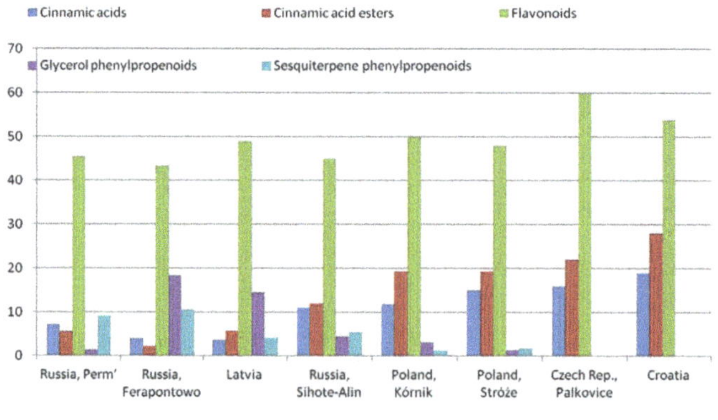

Figure 5.7. Content of individual groups of biologically active components in some poropolis of Eurasia.

This bar graph shows that the main constituents of propolis are flavonoids. The change in the composition of propolis from north to south is also clearly visible. In this direction, the content of cinnamic acids and their esters increases but, remarkably, the "diversity" of the composition decreases. The richest and most diverse are the "northern" propolis; they show the "signals", not only of poplar bud balsams, but also of aspen (glycerol phenylpropenoids) and birch buds (sesquiterpene phenylpropenoids). There is practically nothing of this in the "southern" propolis from the Czech Republic, Crimea, and Croatia. The richer the composition of propolis, the wider the spectrum of its biological activity, which I will talk about in one of the following sections.

In most studies devoted to propolis, extraction is carried out with 70% ethyl alcohol (Greenaway et al, 1988; Bankova et al., 2002; Popova et al., 2007; 2009). In many cases, the sum of the components shown in the graph is less than 100%. When extracting with a water-alcohol mixture from propolis, as the authors cited above do, the sugars contained in it are practically completely extracted. In eight propolis samples, which we subjected to extraction with 70% ethanol (Glinka, 2008), the content of 19 mono- and disaccharides (the main ones were fructose, glucose, and sucrose) was 280–1480 µg/ml and their presence considerably complicates the analysis of phenolic compounds that are much more important from the point of view of the biological properties of propolis. Therefore, for the extraction we chose ether, which practically does not dissolve sugar.

As for the "deficit" in the balance, which can reach 30%, it is explained by the presence of beeswax components in the extracts, including higher alkanes, alkenes, and esters, mainly esters of palmitic and stearic acids. Like sugars, these compounds do not play any significant role in the biological activity of propolis. Therefore, they are not shown on the graph.

Some compounds are present in exudates but are not found in propolis and it is unknown where they disappear to. Conversely, some of the propolis components seem to have fallen from the sky, since they are not present in the exudates of the tree buds in question and there are those that are present in both but in completely different quantities. For example, the contribution of one of the esters of ferulic acid (benzyl ferulate) in extracts of some propolis samples reaches 20–25%, while in poplar exudates it is no more than 2–3% and it does not occur at all in birches.

On this score, two hypotheses suggest themselves. At first glance, the most obvious reason may be the too narrow coverage of the plant precursors of European propolis, perhaps there are more than 6–7. It may also be that not only buds, but other parts, including other plants, also secrete gum substances. It is known, for example, that gum is released from wounds in the trunks of some fruit trees[13]. This hypothesis can be called "botanical".

Previously, it was mentioned that the resin collected by bees, before becoming propolis, is chewed at least twice and mixed with the secretions of the salivary glands of the bees. However, I have not been able to find scientific papers that would the effect of bee saliva on the composition of propolis. Conversely, it has been reported that the secretions of the mandibular gland, considered the true salivary gland, do not contain enzymes, although they are alkaline (Gałuszka, 1998). However, a process called hydrolysis takes place in an alkaline medium. It seems doubtful that enzymes are absent from bee saliva, considering that Polish researchers were able to find two enzymes in propolis: α- and β-amylases (Kaczmarek & Dębowski, 1983). These biocatalysts are well known to degrade complex sugars in which simple sugar molecules, such as glucose and fructose, are linked by glycosidic bonds[14]. Furthermore, there are innumerable plant phenolics combined with sugars by glycosidic bonds, including simple phenols and more complex ones, such as phenolcarboxylic acid, cinnamic acid, and flavonoids. Glycosides are also found in pollen and beebread (Isidorov et al., 2009), as well as raw propolis, and as previously mentioned, the pollen content can reach 10%. It is possible that this may be the reason for the noted discrepancy or that saliva contains not only amylases but other enzymes as well. However, we will leave this for consideration by specialists in the field of physiology and biochemistry of insects and return to the facts already established and try to comprehend them.

There are several conclusions that can be drawn based on the given composition

---

13 *S. Popravko (1982) reported that he and his colleagues failed to find similarities in the composition of propolis and secretions from cracks in the trunks of fruit trees, including plum and apple trees, as well as in the resinous secretions of sunflowers and "... many other similar plants, about which have been mentioned or reported by beekeepers elsewhere as possible sources of bee glue. " True, they used thin layer chromatography in their studies, a method much less perfect than GC-MS and could simply not have noticed the desired compounds if they were contained in these secretions in small quantities.*

14 *Our studies on the composition of sugars in propolis showed a clear preponderance of monosaccharides; the share of fructose and glucose was about 70%, while the share of sucrose was only 5–7% of the fraction. However, what is characteristic in the same fraction, is the 1–3% trehalose, which is formed directly in the bee's body. It is possible, of course, to assume that these sugars appear in propolis due to its "contamination" with honey, but I have serious doubts about this, since we analyzed propolis deposited by bees on a net placed over an area free of frames with honey or the brood.*

of propolis. First of all, the conclusion about its mixed origin, that is, that bees can, if not simultaneously, then at least during one season, collect exudates from different plants. In 80% of the propolis samples we analyzed, a "signal" was found from two or even three tree species. Exudates of the downy birch buds was revealed by the high content of its characteristic flavonoids, sakuranetin and pectolinaringenin, as well as sesquiterpenols and their phenylpropenoids. The presence of glycerol phenylpropenoids indicates that bees also did not ignore aspen buds. Both of these "signals" were present in samples from the north-western region of Russia (Ferapontovo), from the Latvian border region with Russia and from the central part of this country. In all Polish samples, one can see "signals" of birch, aspen, and poplar. The "aspen signal" in these samples weakens from north to south but does not disappear without a trace.

Pure "poplar" propolis originated from the Ukraine and Slovakia. At the same time, with a high degree of probability, it can be assumed that exudates collected there were obtained by bees mainly from the buds of the pyramidal "Italian" form of black poplar (*P. nigra* var. *piramidalis*). This is supported by the ratio of isoferulic and ferulic acids. In Slovak propolis, it was approximately 33:1 and in Ukrainian, it was 26:1. Therefore, it is not necessary to speak only about the purely "poplar" or purely "birch" type of propolis (Popova et al., 2007). The composition of propolis is highly dependent on the contribution of specific resin-giving tree species available to bees and what is attractive to them in a particular area and does not obey a strict typology.

The second conclusion concerns the geographical features of the composition of propolis. It can be seen that the "participation" of black poplar increases from north to south. The fact that northern propolis does not contain compounds characteristic of black poplar is quite understandable as the local climate for this relatively thermophilic plant is not very suitable. Conversely, one should not expect a signal from birch and aspen if propolis comes from the hot climate of the southern European plains. However, the aspen signal appears in propolis collected in the south of the continent in the cooler mountain regions of Bulgaria and Croatia. Significant amounts of phenylpropenoids of glycerol were also found earlier by other researchers in Bulgarian and Swiss propolis and just from the mountainous regions of these countries in which aspen trees grow (Bankova et al., 2002).

These conclusions about the predominantly "mixed" nature of propolis, in the sense that exudates of several plant species are most often its precursors at the same time, and the increase in the contribution of poplar bud exudates when moving from north to south are also confirmed by the results of a study on the composition of commercial propolis extracts with 70% alcohol (Isidorov et al., 2011). They were purchased by me from pharmacies in six different European countries. Below are the concentrations (µg/ml) in solutions of selected groups of components.

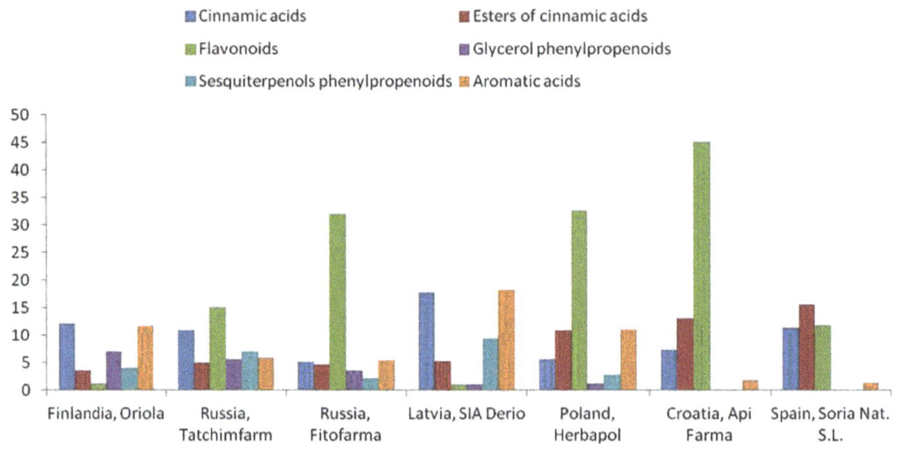

Figure 5.8. The content of certain groups of biologically active components in some commercial preparations of propolis.

Again, we see that the content of phenylpropenoids decreases from north to south, unambiguously indicating downy birch and aspen as donor plants. In contrast, in the same direction, the content of cinnamic acid esters, which are rich in the exudates of poplar buds, increases.

In some regions of Europe, poplar and birch species that are characteristic only of these regions, have buds that are also rich in exudates but their composition is often very different from that typical for the species of these plants. For example, in the mountainous regions of Turkey (Anatolia) and in the Caucasus, a "close relative" of the downy birch (*B. pubescens*), the Litvinov birch (*B. litwinowii*) has a composition of bud exudates that, although close, are not identical. Litvinov's birch secretes some sesquiterpenols and their phenylpropenoids (which will be discussed later), which the northern relative does not have (Isidorov et al., 2015). In Turkey, a thermophilic species of poplar, the Euphrates poplar (*Populus euphratica*), is also often found, in the buds of which 16 phenylpropenoids of glycerol were recorded but not fully identified by Greenawey et al. (1991). Such interspecific features of the composition of exudates should be found and are reflected in the chemical composition of propolis from individual regions. For example, phenylpropenoid glycerides, which are characteristic of it, were found in Turkish propolis collected from where Euphrates poplars grow (Popova et al., 2005). To identify such regional differences in propolis, we studied the compositions of exudates of more than 25 poplar species, including a number of its hybrid forms, and 18 species and varieties of birch, typical for the temperate zone of the Northern Hemisphere, including North America.

There are still other potential plant precursors of propolis to consider, such as the exudates of black alder or horse chestnut. It seems that the rumours about their "propolisity" turned out to be greatly exaggerated. We were unable to detect even traces of fatty acid glycerides characteristic of black alder buds in any propolis sample (Isidorov et al., 2016). In one of the publications (Nyeko et al., 2002), it was reported that in Uganda, bees use alder foliage to prepare propolis so actively that it causes defoliation of these trees. I have no reason to dispute this report, since I myself have not been to Uganda, I have not seen alder, and I have not analyzed propolis from this country. However, we can say with almost complete certainty that in our latitudes the bees of the species *Apis mellifera*, alder are of little interest.

Horse chestnut exudates contain only small amounts of flavonoids; their total content is 2.5%. The buds of this plant are characterized by the presence of a series of six 3-hydroxy acids C_{12}–C_{22} (Isidorov et al., 2016). We registered small amounts of two of them, 3-hydroxypalmitic and 3-hydroxystearic acid, in several samples of propolis from the northern regions of Russia, where, however, there is no chance of meeting this plant. Small amounts of just these two 3-hydroxy acids were found in aspen bud exudates.

It is unknown if in a given area there were a sufficient number of different types of "resin-giving" trees, the bees would show a preference for these sources. I hypothesize that bees are selective and in support of this, I will cite two of my own observations.

First observation. In the summer and fall of 2010, I sampled propolis twice in an apiary near the village of Tirza, in the north-east of Latvia. I collected samples from three hives and looked closely at the surrounding vegetation. I observed that the hives were under linden and maple trees but about 25–30 meters away from them, five or six spreading poplars were growing, the age of which were determined by eye at 50–60 years old. This was located in the middle of the field, behind which, about 350–400 meters from the apiary, there is a mixed forest of pine, spruce, birch, aspen, and various shrubs. Of the trees present in the apiary, I assumed a priori that it must be poplar being utilized by the bees for propolis. Indeed, of all the trees growing nearby, only the poplar very generously offers its resin to the bees, and the bee needs to fly to birches and aspens and return with a considerable load. As it turned out, the composition of propolis was typically birch-aspen. It did not even contain traces of compounds characteristic of poplar, such as methylbutenyl caffeates, chrysin, pinobanksin, and chalcones. In addition, the local propolis was more similar to that collected in Ferapontovo, about 650 km northeast of Tirza, than from an area 150 km southwest in the same Latvia (Ogre).

Upon closer inspection, it turned out that the foliage of these trees is different in shape from the foliage of the black poplar. Chemical analysis of the exudates of the buds of these trees showed that it is a North American balsamic poplar (Populus balsamifera), as evidenced by the uniquely high content of dihydrochalcones (see section 5.5), which is characteristic only of this species. This feature of balsamic poplar was noted earlier (Greenaway et al., 1991). According to the testimony of local residents, poplars were planted in the territory of the former forestry and the planting material came from a plant nursery that offered various exotic species for the area. However, the foreign origin of trees is difficult to explain the apparent neglect of them by bees. Maybe the reason lies in the high content of dihydrochalcones, which are atypical for the native black poplar and also unusual for our bees; in the ether extract of P. balsamifera buds exudate, according to my data, it was 42.2%.

Second observation. In September of the same year, propolis was collected from hives in an apiary located near Bialystok in north-eastern Poland. The hives on it were located at the very edge of a mixed forest, consisting mainly of birch, aspen, and pine. A walk around the nearest surroundings showed that birch was represented here by only one species, the silver birch (*Betula pendula*), which was confirmed by chemical analysis of the collected buds. It is clear that propolis should be birch-aspen and so it turned out, in all three samples, the signals of aspen and birch were perfectly clear but only with a slight correction. The birch signal did not come from silver birch but from downy birch (*B. pubescens*), which was not near the apiary. There was not even a hint of silver birch exudates in this propolis. When considering the propolis compositions, none of the samples contained triterpenoids characteristic of silver birch, such as the aforementioned dammaradien-3-one and dipterocarpol.

Downy birch, from which the bees collected exudates, was found at a distance of about 250–300 meters from the hives. The bees flew there, ignoring the resinous secretions on the buds of the silver birches, under which the apiary was located. This kind of selectivity was noted somewhat later by American authors (Wilson et al., 2013). They observed that the bees collected the resinous exudates of the buds of *Populus balsamifera* and *P. deltoides* but ignored other poplar species growing nearby. They were also not interested in the abundant discharge from the buds and cracks in the bark of pine and spruce.

From all that has been said, the third, and perhaps the most important, conclusion can be drawn that not everything that sticks, is available in large quantities, and grows nearby, attracts bees. For them, it is not only the consistency of the collected material, which allows it to be used for repair and construction needs, that is important but primarily the special properties of its chemical compounds. Since the main function of propolis in the bee colony is to suppress the development of pathogenic microbes (Evans & Spivak, 2010; Simone-Finstorm & Spivak, 2010), bees should be interested in compounds with antiseptic action. The second hypothesis of selectivity consists in the possible presence of exudates with compounds that repel or are toxic to bees.

To test the first hypothesis, we carried out comparative studies of the antimicrobial activity of bud exudates of seven woody plant species and seven propolis samples of different types and origins (Isidorov et al., 2016). The tests were carried out on microbes found on honey plants and pathogenic for bees. The results of the study were unexpected. First, all propolis samples, despite large differences in chemical composition, demonstrated similar activity. Secondly, exudates of discriminated plant species, such as silver birch, black alder, and pine, had fairly high antimicrobial activity.

Thus, our microbiological studies have not provided an unambiguous answer to the question of why bees ignore or avoid certain plant species. A similar result was previously obtained by American authors (Wilson et al., 2013), who studied the behavior of bees collecting resins of various plants and their effect on the dangerous pathogen of bees, the bacterium *Paenibacillus larvae*. It was found that bees collect poplar (*Populus deltoides*) exudates, which is characterized by moderate antimicrobial activity against this pathogen (MIC 0.175 mg/ml) but does not collect much more active (MIC 0.05–0.06 mg/ml) resinous exudates from spruce *Picea glauca* and pines *Pinus banksiana* and *P. ponderosa*.

Since we are again talking about conifers as probable plant precursors of propolis, I will emphasize once again that in propolis of the boreal and subboreal belts, that is, in the areas of Scots pine (P. sylvestris) growth, its signals are practically not found. They are diterpenoids called resin acids. However, they have been found in propolis from Greece, Cyprus, and the Aegean islands, which have neither birches nor poplars (Kalogeropoulos et al., 2009; Melliou & Chinou, 2004; Popova et al., 2009). The plant precursors of resin acids in these areas are conifers of the Cupressaceae and Pinaceae families, such as Pinus pinea. P. nigra, P. halepensis, and P. brutia. Bees are simply forced to use this material for lack of anything else.

As for the second hypothesis (about compounds with toxic or deterrent action), there are currently no serious candidates for this role. The exudates of the buds of silver birch contain noticeable amounts of the flavonoid, catechin, which is practically absent in the exudate of downy birch but its toxicity has not been tested in bees.

I will allow myself to express my own explanation of the different attitudes of bees to the secretions of these two species of birch. The main difference in the chemical composition of exudates concerns the content of flavonoids and other aromatic compounds. In the case of downy birch, they account for about 64% of the mass of exudates, while in silver birch, they are less than 1%; there is also little in the secretions of the buds of horse chestnut, black alder, and pine. Meanwhile, all aromatic compounds absorb radiation in the near ultraviolet range, that is, in the range that bees can easily distinguish. For them, droplets of exudates on the buds of downy birch and poplars shine like light bulbs that decorate Christmas trees. At the same time, buds devoid of aromatic compounds are invisible to bees. However, this hypothesis has not yet been tested.

The antibiotic properties of propolis has long attracted human attention. However, studies carried out in recent decades have made it possible to discover many new beneficial properties of propolis. Thanks to these discoveries, it finds more and more widespread recognition, even in official medicine. In due time we will take a closer look at these miraculous substances.

5.4. What does propolis smell like?

Everyone who has dealt with "raw" propolis remembers its characteristic smell. It has been reported to scent of flowers, cinnamon, and various kinds of balms. However, it has never been reported that the smell is unpleasant. The smell is determined by the presence of a range of volatile organic compounds (VOCs) in propolis. It is surprising to note that the study of their composition until now has received incomparably less attention than low-volatile, extractive compounds. The same, incidentally, applies to plant exudates, the volatile secretions of which have been studied very poorly. Perhaps this is due to the fact that until recently, the determination of the composition of VOCs was associated with long-term painstaking work to isolate them from propolis, and, moreover, requires special equipment.

English scientists became interested in one of the first volatile secretions of propolis a little over 30 years ago. They placed 45 g of propolis in a closed vessel and slowly purged it with gas for 16 hours and the evaporated VOCs were captured with a special sorbent (Greenaway et al., 1990). By subsequently heating the sorbent in a special device, called a thermal desorber (which is very expensive), and purging it with an inert gas, VOCs are transferred to a gas chromatographic column and analyzed by GC-MS. Of the 29 registered compounds (24 of them were identified), 12 turned out to be esters, mainly acetates, which are usually found in various essences with a fruity smell. The rest of the VOCs of the sample under study were lower alcohols, carbonyl compounds, and acids, including benzoic acid.

Croatian researchers (Borćič et al., 1996) used a different technique, called steam distillation, which is widely used to obtain essential oils from plants, to isolate VOCs from two propolis samples collected in different parts of the country. In one of the samples, it was possible to identify 13, and in the other, 16 compounds, the composition of which was different from that reported by the English authors. The VOCs of Croatian propolis included mainly terpene compounds, which give the propolis a floral scent. Interestingly, naphthalene was found to be the main component in the volatile emissions of one of the propolis samples. The authors did not take responsibility for identifying the source of this compound but cited evidence from local beekeepers that in areas poor in arboreal vegetation, bees may collect "unexpected materials", such as asphalt.

Greek researchers reported a significantly higher VOC "catch" in five samples from different parts of the country (Melliou et al., 2007). They also resorted to long-term steam distillation and used large amounts of starting material in their experiments, 1 kg of each of the samples; this extravagance is justified by the fact that the authors set themselves the goal of determining the antimicrobial activity of VOCs, which requires rather large amounts of distillate. As a result of distillate analysis by GC-MS, 42–56 compounds were identified in each of the samples, the composition of which only partially overlapped; of 94, only 16 compounds were discovered in all five propolis samples. Another 10 components were present simultaneously in four samples. The overwhelming majority of these common compounds turned out to be terpenes and their derivatives (terpenoids).

It is safe to say that the authors of even this work, who used such large quantities of raw propolis, were unable to discover all the volatile components it contained. Everyone who has been involved in the distillation of plant material with water vapour remembers what aroma is near the installation. This aroma consists of the most volatile of all VOCs that have escaped along with the steam and, therefore, have been irretrievably lost for further research.

Much more economical in terms of both material and time consumption is a relatively new method of sample preparation for analysis, called solid phase microextraction (SPME). The fundamentals of this method were developed by Janusz Pawliszyn in the early 1990s at the University of Waterloo in Canada. The essence of the method lies in the fact that a quartz fiber is introduced into the object under study, placed in a closed vessel, using a special device resembling a syringe with a retractable needle. This fiber is coated with a sorbent layer (7–30 µm thick), in which volatiles are trapped. The object of investigation can be liquid, for example, drinking or wastewater, or it can be solid, but then the fiber is introduced into the gas phase above the sample. The volatile compounds emitted by it are captured by the sorbent on the fiber surface. After the completion of the stage of trapping VOCs (their concentration), the fiber is introduced using the same syringe into the heated evaporation chamber of the chromatograph. The evaporated components are taken up by a current of an inert carrier gas and transferred to a chromatographic column. Furthermore, everything is the same as in the analysis of liquid extracts, including separation of components, their identification, and interpretation of the analysis.

It seems very attractive to use SPME to establish the origin of propolis, that is, its plant precursors. This is primarily due to significantly lower costs of time and material in comparison with the method described above. The SPME technique reduces the analysis time 3.5–4 times and the amount of propolis required for it is only about 1 g.

Naturally, to solve this problem, it is also necessary to know the chemical composition of the volatile secretions of tree buds, from which bees can collect exudates. Searches in the scientific literature have led nowhere; researchers, including the author of this book, intensively studied the volatile secretions of tree foliage and flowers but so far have not been interested in the smell of buds. The SPME technique combined with gas chromatography is so sensitive that for our purposes only 10–15 buds with exudates were enough. Earlier studies allowed us to restrict ourselves to the buds of only those trees in our climatic zone, the "propolisity" of which can be said with complete confidence. As a result of experiments, 102 compounds were identified in the volatile excretions of poplars and downy birch, and 124 compounds were identified in the excretions of propolis (Grzech, 2011). The chromatograms below indicate how rich the composition of the volatile secretions of the buds are, which determines their specific odour.

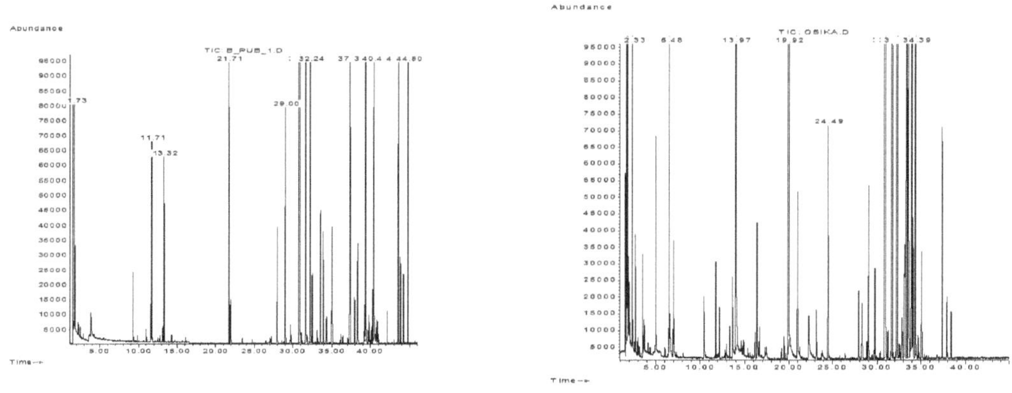

Figure 5.9. Chromatograms of volatiles from the buds of downy birch (left) and aspen buds (right).

As expected, the list of propolis VOCs has significantly expanded in comparison with that given by Greek researchers (Melliou et al., 2007), mainly due to the most volatile compounds that were "lost" during steam distillation. These are light alcohols, carbonyl compounds, and ethers. Some of them refer to nonspecific metabolites, common to all living organisms (ethanol, acetone, ethyl acetate), but some components may specifically indicate the type of plant. The indicator group for poplar buds is formed by unsaturated substances with the same carbon skeleton (Fig 5.10).

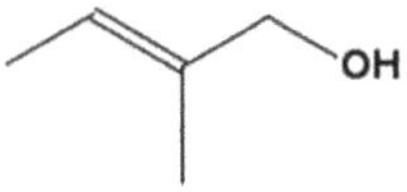

1

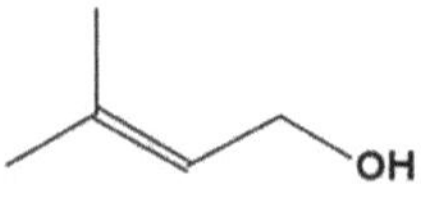

2

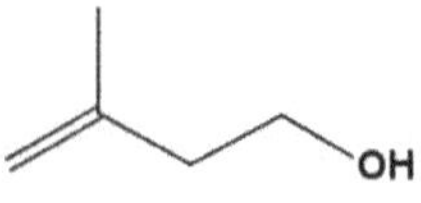

3

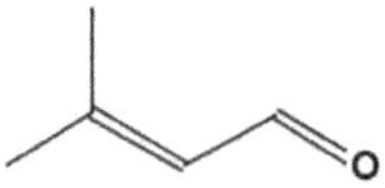

4

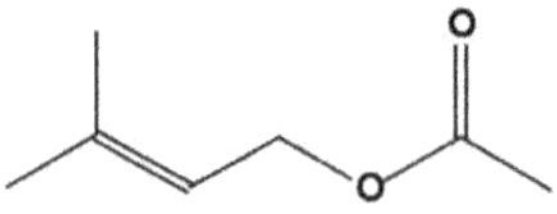

5

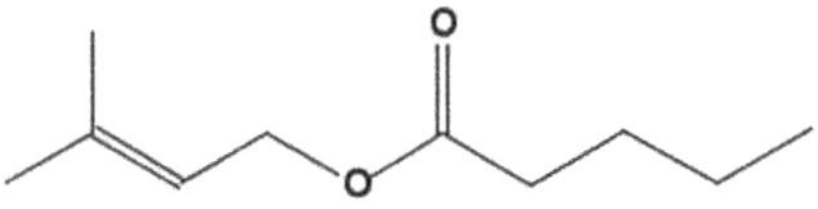

6

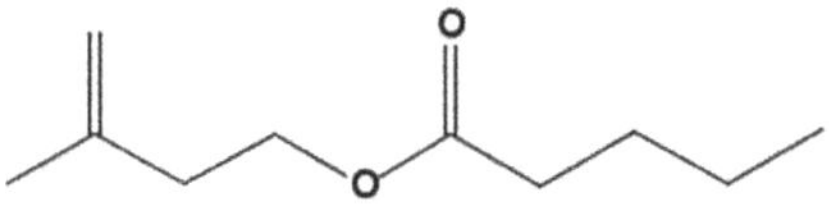

7

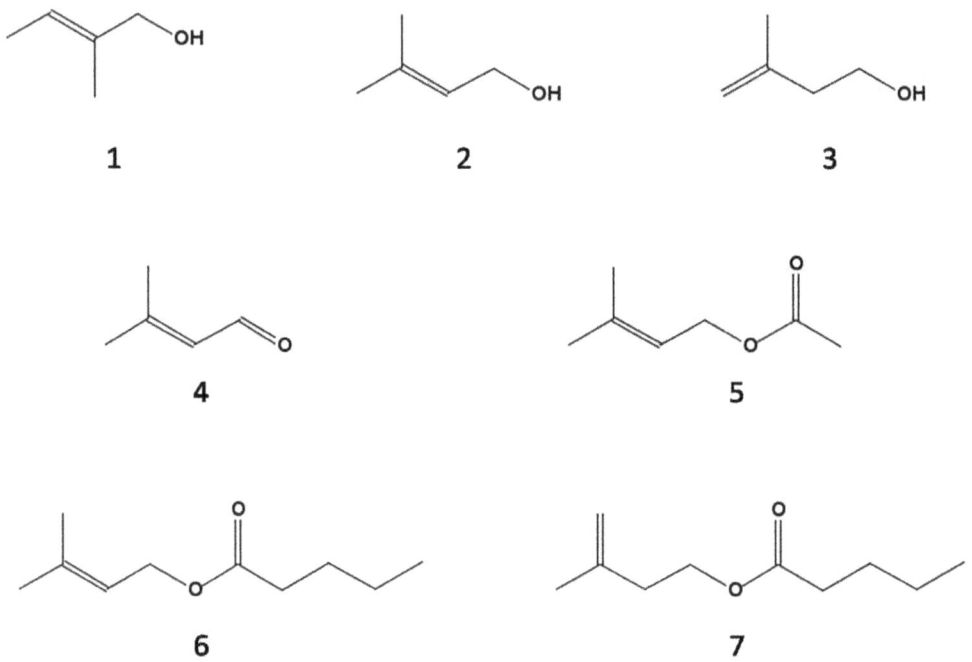

Figure 5.10. Formulas of compounds typical for the volatile secretions of poplar buds. 1 – 2-methyl-2-buten-1-ol, 2 – 3-methyl-2-buten-1-ol (prenol), 3 – 3-methyl-3-buten-1-ol, 4 – 3-methyl-2-butenal (prenal), 5 – 3-methyl-2-butenol acetate (prenyl acetate), 6 – 3-methyl-2-butenyl valerate, 7 - 3-methyl-3-butenyl valerate.

Some of these compounds were previously found in propolis by British researchers (Greenaway et al., 1990). We have already been met with similar structures in the ether extracts of poplar buds and in many propolis samples; there are many (as many as 17) esters of unsaturated alcohols $C_5H_{10}O$ and cinnamic acids, such as *p*-hydroxycinnamic (*p*-coumaric), ferulic, isoferulic, and caffeic acids. The biological activity of these compounds will be discussed in the next section. Below is the composition (%) of the volatile excretions of the buds and some of the propolis we studied.

Table 5.2. Average VOCs composition (%) of some plant buds and propolis

Groups of compounds	*Poplar*			*Birch, B. pubescens*	*Propolis**		
	P. nigra	*P. nigra var. piramidalis*	*P. tremula*		*A*	*B*	*C*
Carbonyl compounds	2.4	2.2	3.9	0.3	1.4	3.6	3.8
including:							
- 3-methyl-2-butenal	0.02	0.7	-	-	-	1.4	0.2
- benzaldehyde	1.4	0.1	0.2	-	0.8	0.7	0.1
- salicyl aldehyde	0.8	0.2	2.8	-	-	-	-
- 6-methyl-5-hepten-2-one	-	-	-	0.2	0.02	-	0.5
Alcohols	11.5	12.3	8.9	0.3	2.6	19.5	1.4
including:							
- 3-methyl-3-buten-1-ol	1.7	3.5	-	-	-	4.3	-
- 3-methyl-2-buten-1-ol	1.0	5.4	-	-	-	6.2	0.2
- benzyl alcohol	4.2	0.2	0.3	-	2.1	1.1	-
- 2-phenyletanol	1.8	1.3	trace	-	0.5	7.4	-
Esters	5.8	5.0	3.5	2.5	trace	22.2	1.1
including:							
- 3-methyl-2-buten-1-ol acetate	0.7	2.9	-	-	-	13.9	-
- methyl salicylate	3.6	1.4	0.6	-	-	-	-
- methyl benzoate	0.5	-	0.5	-	-	-	-
- ethyl benzoate	-	-	5.3	-	-	-	-
Monoterpenoids	10.1	16.2	1.1	1.5	4.4	3.3	6.15
including:							
- 1,8-cineol	-	3.2	-	-	trace	0.3	-

Sesquiterpenoids	54.2	42.4	71.5	85.0	56.5	4.1	81.9
including:							
- β-caryophyllene	19.2	16.1	40.5	41.7	1.7	0.5	12.4
- α-humulene	2.2	2.0	10.8	9.5	0.7	0.3	2.8
- bulnesene	-	0.9	2.1	-	-	0.3	-
- α-guaene	-	trace	11.9	trace	0.6	-	-
- birkenal	-	-	-	22.5	24.8	-	21.8
- 6-hydroxy-β-caryophyllene	-	-	-	3.5	6.1	-	12.9
- 14-hydroxy-β-caryophyllene	-	-	-	2.2	0.7	-	3.5
- 6-hydroxy-β-caryophyllene acetate	-	-	-	3.9	0.9	-	0.3
- 14-hydroxy-β-caryophyllene acetate	-	-	-	8.1	9.9	-	14.4
α-,β-, γ-eudesmol	5.1	-	-	-	-	0.9	-

A – Ferapontowo, Russia; B –Poltava region, Ukraine; C – Białystok, Poland

The composition of the volatile components makes it easy to distinguish the buds of poplars from the buds of downy birch; the latter emit large amounts of terpene compounds, such as birkenal, hydroxycaryophyllenes, and their acetates (Fig. 5.11)which are not among the volatile substances of silver birch. Therefore, the microextraction technique can be successfully applied to the discovery of the components of poplar and downy birch exudates in propolis, that is, to establish the "type" of propolis (Isidorov et al., 2014). Moreover, these sesquiterpenoids were found in the volatile secretions of propolis from Ferapontovo and Bialystok, in the ether extracts of which the "birch signal" was very expressive.

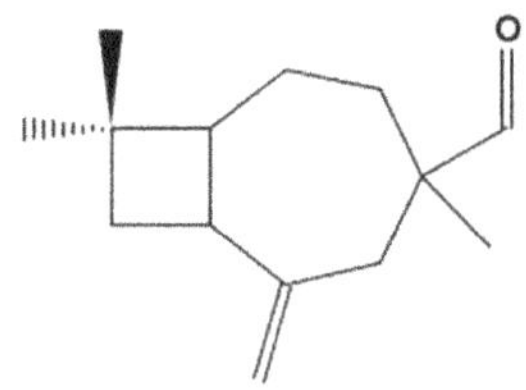

1

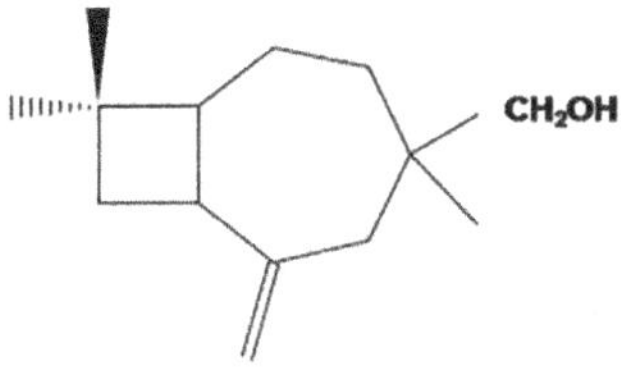

2

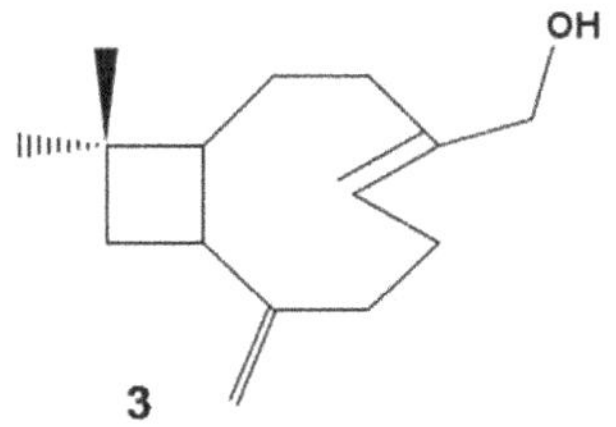

3

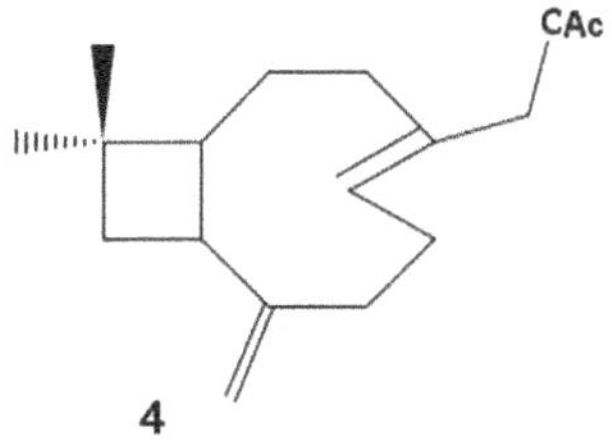

4

Figure 5.11. Sesquiterpenoids, typical for resinous secretions of the buds of downy birch and for "birch type" propolis. 1 - birkenal, 2 - birkenol, 3 - 14-hydroxy-β-caryophyllene, 4 - 14-hydroxy-β-caryophyllene acetate.

The most characteristic compounds for aspen, glycerides of cinnamic acids, are non-volatile. However, the combination of salicylic aldehyde, benzyl benzoate, and large amounts of the terpene alcohol α-guaiene will also help to identify the aspen signal in propolis. It is also difficult to draw a conclusion, with the help of SPME, about which of the two forms of black poplar (*P. nigra* or *P. nigra* var. *piramidalis*) was the plant precursor of the studied propolis. However, this is not the most important thing. More importantly, the inner atmosphere of the bee's home is full of many volatile organic compounds and it is unknown what effect they have on the life of bees.

Interestingly, the composition of the VOCs of the buds and propolis do not completely coincide; propolis releases a fairly large number of compounds that are not found among the volatile compounds of the buds of "resin-giving" trees. These are lower aliphatic acids, from formic to octanoic (most of them are acetic acid), as well as esters of these acids, mainly acetates. Again, the question arises about the origin of these additional compounds. In principle, acetic, butyric, and other lower acids, as well as their esters (acetate, butyrate, and others), can be formed during the microbiological oxidation of sugars. However, it is difficult to believe that any bacteria or fungi can survive in propolis, which has a strong antimicrobial effect. Again, the assumption is that additional substances are formed in enzymatic processes. Thus, it is likely that the source of enzymes in propolis are secretions from the salivary glands added by bees to the exudates of plants.

5.5. What does propolis treat and how?

I hope that I have been able to make a convincing case that bees carefully choose the plant material for the preparation of propolis. There is no doubt that this choice is determined by the chemical composition of its botanical precursors. What chemical compounds seem preferable to them, and what useful properties do they have? We thought that the direct answer to this question was given by an example with birches; downy birch is very popular with bees, but silver birch is somehow not attractive. The reason for this seemed clear as the exudates of the buds of downy birch contain large amounts of polyphenols and the silver one does not.

Polyphenols are rich not only in the buds of downy birch but also in all types of poplars; however, their content is low in the buds of horse chestnut and black alder. It is with these compounds that the main properties of propolis are associated, including its antiseptic effect. This characteristic of propolis has been known since ancient times, but scientific confirmation was obtained relatively recently. A pioneer in the study of its antimicrobial properties was V.P. Kivalkina, who first conducted the appropriate microbiological tests in 1947 at the Kazan Veterinary Institute (Kivalkina, 1948). Therefore, we will focus our attention specifically on polyphenolic compounds, primarily on flavonoids.

Flavonoids belong to the so-called group of C6–C3–C6 compounds, since their skeleton contains two benzene rings of six carbon atoms each interconnected by a heterocycle, which includes three more carbon atoms. Depending on the degree of oxidation of the heterocyclic moiety, flavonoids can be categorized into six main groups: flavones, flavanones, flavonols, flavanols (also called catechins), anthocyanidins, and leukoanthocyanidins. Of these six groups, the first four are of particular interest, since they are widely represented in propolis.

From a biochemical point of view, another group of polyphenols, chalcones, is closely related to flavonoids. These are unsaturated ketones with two aromatic rings connected by a linear chain of three carbon atoms, that is, they also belong to C6–C3–C6 compounds. Chalcones are easy to cyclise, resulting in the formation of flavanones. Conversely, the opening of the pyrane heterocycle of flavanones under the action of certain agents can lead to the formation of chalcones and these processes are, in principle, reversible (Fig. 5.12).

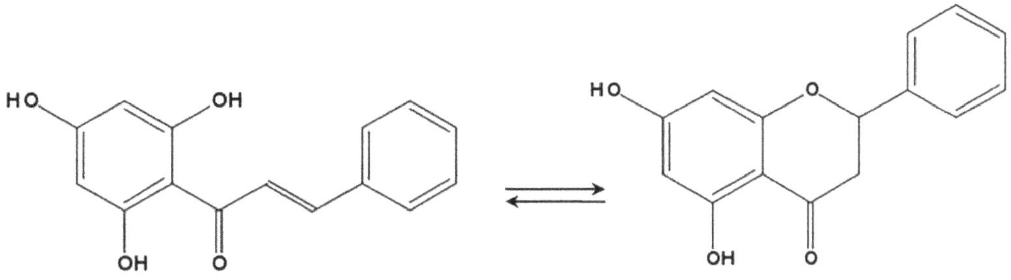

1 2

Figure 5.12. Mutual reversible conversion of pinocembrin chalcone and pinocembrin flavonoid.

In plant materials including bud exudates, both forms, flavanones and their chalcones, are often found. In exudates, products of partial reduction of chalcones, dihydrochalcones, are also often found. Particularly high in their content (up to 50% of the total mass of exudate), the excretions of buds of balsamic poplar (*P. balsamifera*) are distinguished. British authors suggested that the content of chalcones and dihydrochalcones is so high due to the low activity of the chalcone isomerase enzyme in balsamic poplar (Greenaway et al., 1989). Under the action of this enzyme, the cyclization of chalcones occurs.

In the plant world, there are many hundreds of natural flavonoids (some literary sources say that there are at least 2000 of them) and this diversity is achieved due to the fact that in rings A and B, groups –OH and –OCH_3 are in different positions and combinations. Of all the various compounds in European propolis, 64 were found and are listed below.

Table 5.3. The main groups of flavonoids and chalcones in propolis

Group	Chemical name	Synonim
Flavone 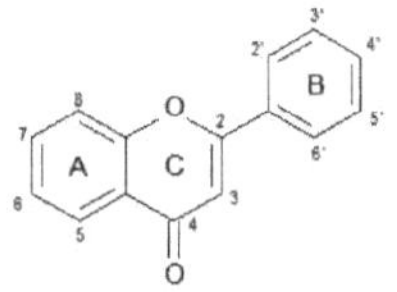	5-Hydroxy-7-methoxyflavone	Tectochrysin
	5,7-Dihydroxyflavone	Chryzin
	5,7-Dihydroxy-4'-metoxyflavone	Acacetin (linarigenin)
	5-Hydroxy-4',7-dimethoxyflavone	4'-Methylgenkwanin (4,7-dimethylapigenin)
	5,7-Dihydroxy-4',6-dimethoxyflavone	Pectolinaringenin
	5,7,4'-Trihydroxyflavone	Apigenin
	5,7-Dihydroxy-3-methoxyflavone	3-Methylgalangin
	3,5,7,4'-Tetrahydroxyflavone	Kaempferol
	3,5,7-Trihydroxyflavone	Galangin
	3,5,7-Trihydroxy-4'-methoxyflavone	Kaempferide
	3,5-Dihydroxy-4',7-dimethoxyflavone	-
	5,7-Dihydroxy-3,4'-dimethoxyflavone	-
	3,4',5-Trihydroxy-7-methoxyflavone	Rhamnocitrin
	5,7,4'-Trihydroxy-3-methoxyflavone	3-Methylkaempferol
	3,3',4',5-Tetrahydroxy-7-methoxyflavone	7-Methylquercetin (rhamnetin)
	3,3',5,7-Tetrahydroxy-4'-methoxyflavone	4'-Metylo quercetin
	4',5,7-Trihydroxy-3,3'-dimethoxyflavone	3,3'-Dimethylquercetin
	3,3',4',5,7-Pentahydroxyflavone	Quercetin
	3,3',4', 5,5',7-Hexahydroxyflavone	Myricetin
	4',5-Dihydroxy-7-methoxyflavone	Genkwanin

Flavanone 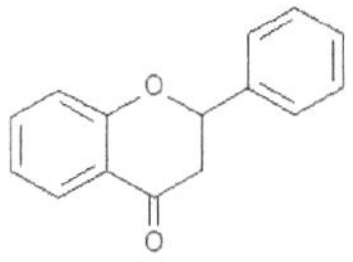	5,7-Dihydroxyflavanone	Pinocembrin
	5-Hydroxy-7-methoxyflavanone	Pinostrobin
	4',5-Dihydroxy-7-methoxyflavanone	Sakuranetin
	3,5-Dihydroxy-7-methoxyflavanone	Izalpinin
	5,7-Dihydroxy-4'-methoxyflavanone	Isosakuranetin
	3,5,7-Trihydroxyflavanone	Pinobanksin
	3,7-Dihydroxy-5-methoxyflavanone	Pinobanksin 5-methyl ether
	5,7-Dihydroxy-3-acetylflavanone	3-Acetyl pinobanksin
	5,7-Dihydroxy-3-propanoylflavanone	3-Propionyl pinobanksin
	5,7- Dihydroxy-3-isobutanoylflavanone	3-Izobutanoyl pinobanksin
	5,7- Dihydroxy-3-butanoylflavanone	3-Butanoyl pinobanksin
	5,7- Dihydroxy-3-pentanoylflavanone	3-Pentanoyl pinobanksin
	5,7- Dihydroxy-3-pentenoylflavanone	3-Pentenoylo pinobanksin
	5,7- Dihydroxy-3-hexanoylflavanone	3-Heksanoylo pinobanksin
	5,7,4'-Trihydroxy-3-metoxyflavanone	Homoeriodictiol
	7-Hydroxy-5-methoxyflavanone	Alpinetin
	5-Hydroxy-4',7-dimethoxyflavanone	4'-Methoxysakuranetin
	5,7,4-Trihydroxy-4-metoxyflavanone	-
	3',5,7-Trihydroxy-4'-methoksyflavanone	Hesperetin
	4',5,7-Trihydroxyflavanone	Naringenin
	3,5,7-Trihydroxy-4'-methoxyflavanone	-
	3,5,4'-Trihydroxy-7-methoxyflavanone	-
	3,4',5,7-Tetrahydroxyflavanone	Aromadendrin
	5,7-Dihydroxy-3,4'-dimethoxyflavanone	Ermanin, 3,4-dimethylkaempferol
	5,7,3',4'-Tetrahydroksyflawanon	Luteolin
Flavanol 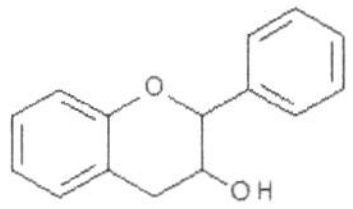	3,3',4',5,7-Pentahydroksyflawanol (trans-)	(+)-Catechin
	3,3',4',5,7-Pentahydroksyflawanol (cis-)	(-)-Epicatechin

Chalcone	2’,6’-Dihydroxy-4’-methoxychalcone	Pinostrobin chalcone
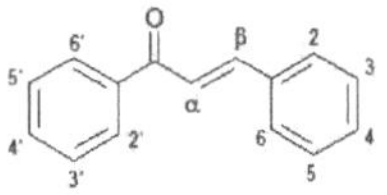	2’,4’,6’-Trihydroksychalcone	Pinocembrin chalcone
	2’,4’-Dihydroksy-6’-methoxychalcone	Alpinetin chalcone
	2’,6’,α-Trihydroksy-4’-methoxychalcone	-
	2’,6’-Dihydroksy-4,4’-dimethoxychalcone	-
	2’,4’,6’-Trihydroksy-4-methoxychalcone	Isosakuranetin chalcone
	2’,4’,6-Trihydroksy-4’-methoxychalcone	Sakuranetin chalcone
	2’,4’,6’,4-Tetrahydroxychalcone	Naringenin chalcone
	3,5,7-Trihydroksy-44-methoxychalcone	-
Dihydrochalcone	2’,6’-Dihydroxy-4’-methoxydihydrochalcone	-
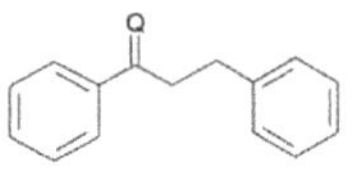	2’,4’,6’-Trihydroxydihydrochalcone	-
	2’,4’,6-Trihydroxy-4-methoxydihydrochalcone	-
	2’,6’-Dihydroxy-4,4’-dimehoxydihydrochalcone	-
	2’,4’,6’-Trihydroxy-4-methoxydihydrochalcone	-
	2’,4,6’-Trihydroxy-4’-methoxydihydrochalcone	-
	2’,4’,6’,4-Tetrahydroxydihydrochalcone	Phloretin

The techniques of chemical analysis are constantly being improved, so it can be expected that this list will be replenished with newly discovered substances of this class in the near future.

The total content of flavonoids in propolis of different origin can vary within a fairly wide range. Japanese researchers (Kumazawa et al., 2004) studied 16 propolis samples from 14 countries from all continents of the globe; although, interestingly, there were no propolis samples from Japan among them. They found that in the dry residue after extraction with ethanol and removal of the solvent (this material is usually denoted by the abbreviation EEP), the total content of ten flavonoids varied more than three times and was in the range of 50–176 mg/g EEP (on average, 124±41 mg/g EEP). Only one sample, originating from Thailand, had an extremely low content of these compounds (2.5±0.8 mg/g EEP). In most propolis, the main flavonoids were chrysin and pinocembrin, followed by pinobanksin 3-acetate.

Italian scientists also conducted a comparative study of flavonoids in alcoholic propolis extracts from seven countries. The technique used by them (HPLC-MS), due to the low separation power of liquid chromatography, made it possible to identify only eight flavonoids. In six samples, their total concentrations were in the range of 176–346 μg/ml and pinocembrin was present in the highest amounts (from

40 to 68% of the total content) in all samples. It was followed by naringenin and galangin (Volpi & Bergonzini, 2006). None of the eight flavonoids were found in propolis from Kenya.

The graph below (Fig. 5.13) shows the content of flavonoids and other phenolic compounds in alcohol (70% ethanol) extracts of eight propolis samples from Russia, Poland, and Latvia.

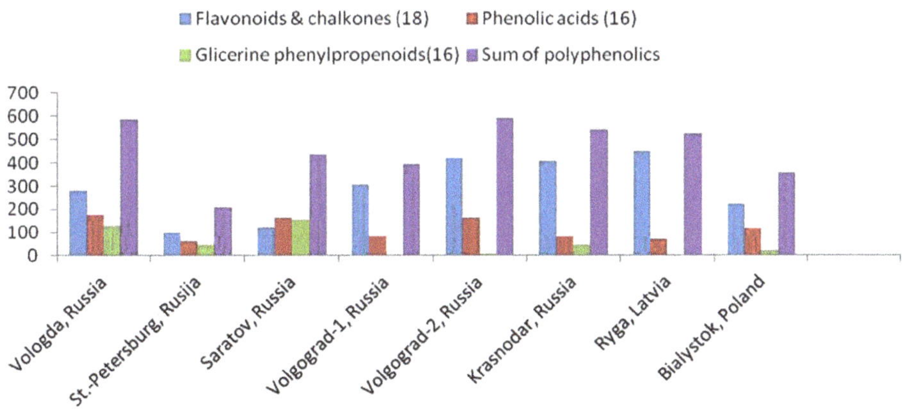

Figure 5.13. The total content (mg/g EET) of phenolic compounds in the alcoholic extract of propolis from the mid-latitude regions of the Northern Hemisphere

The total content of 18 flavonoids and chalcones also varied greatly among samples within the range of 99–450 mg/g EEP and 288±134 mg/g EEP, respectively. In samples from the northern regions of Russia (Vologda and Leningrad regions), as well as in Polish propolis, sakuranetin was found in the largest quantities. Pinocembrin and chrysin dominated in propolis from the southern regions of Russia. In the most flavonoid-rich sample from southwest Latvia, the main components were pinobanksin 3-acetate and pinocembrin. These results are quite consistent with the geographic features of propolis composition noted above.

The number of scientific publications devoted to the antimicrobial properties of propolis and the flavonoids it contains are enormous. Most of them emphasize that propolis is active against both gram-positive and gram-negative bacteria at flavonoid concentrations of less than 10 mg/ml.

Perhaps one of the broadest research subjects in terms of coverage was the work of Italian scientists who studied the effect of propolis on 320 strains of microorganisms (Drago et al., 2000). The results showed that it had pronounced antimicrobial activity against most isolates. Pathogens such as *Streptococcus pneumoniae*, *Klebsiella pneumoniae*, and *Haemophilus influenzae* were particularly sensitive. However, enterobacteria turned out to be quite resistant, including the notorious *Escherichia coli*, which recently caused a real panic in the countries of Western and Central Europe. The main conclusion of the authors was that propolis has pronounced bacteriostatic properties and exhibits a bactericidal effect only at high concentrations of the extract.

Furthermore, high concentrations, that is, to some extent an overdose of propolis flavonoids, obviously does not threaten human health. Toxicological studies carried out as far back as the 1980s have shown very low levels of toxicity for these plant polyphenols. The so-called half-lethal dose (LD50) is used as a measure of toxicity, expressed in mg of the test substance per 1 kg of body weight. The higher the LD50 value, the safer the substance. Flavonoids are characterized by surprisingly high LD50 values, which range from 8–40 mg/kg. The LD50 for EEP in mouse experiments was 7.34 g/kg (Havsteen, 1983). Therefore, the flavonoids identified in propolis belong to the group of compounds generally recognized as safe (GRAS). It is also important that flavonoids are metabolized without the formation of toxic products that could accumulate in the body to a level hazardous to human health. Therefore, propolis can be used not only as an external but also as an internal agent.

Long-term use of propolis, apparently, also does not have negative consequences, including the occurrence of dysbiosis. At least they could not be detected in experimental animals that received high daily doses of propolis (2.5–4.0 g/kg) for three months. For one of the flavonoids, quercetin, a carcinogenic effect was established in experiments on rodents but pure quercetin is not typically consumed; moreover, this flavone is not among the main components of propolis. Toxicologists, who are very careful when it comes to transferring the results obtained in animal experiments to humans, use a safety factor equal to 1000. Therefore, if the daily dose that does not cause any negative effect in mice is, for example, 4000 mg/kg, then for humans it is estimated at 4 mg/kg. That is, with a body weight of 70 kg, an absolutely safe dose for internal administration is approximately 280 mg per day. The recommended daily dose of propolis (in powder form) in the United States, where it is widely used as a dietary supplement, is 200 mg (Burdock, 1998). However, according to some reports, a diet rich in fruits and vegetables may contain a daily dose of flavonoids up to 650 mg (Liu, 2004).

Among the relatively recent discoveries on the biological properties of flavonoids, their anti-inflammatory effect deserves special attention. Inflammation is a normal biological process as the body's reaction to all kinds of tissue damage, microbiological infection, or chemical irritation. In the damaged area, inflammation is expressed in the intake of immune cells from the blood vessels and the release of substances that can form active particles that attack the pathogenic factor. The normal inflammatory process proceeds fairly quickly and without any negative consequences. However, if inflammation becomes chronic, it can lead to chronic disease. This is due to the fact that inflamed cells generate reactive oxygen species (hydroxyl radicals $HO^{\bullet}$, superoxide radicals $^{\bullet}O_2^-$, singlet oxygen $^1O_2$) and nitric oxide, NO, which is also a free radical.

An excess of these oxidants leads to the development of a variety of diseases, including the main cause of neurodegeneration, damage to nerve cells. It has been established that the cause of such formidable neurodegenerative diseases, such as Alzheimer's and Parkinson's, is the destruction of cholinergic and dopaminergic neurons of the central nervous system under the action of oxidants. Oxidative stress also underlies the development of many forms of cancer, rheumatoid arthritis, and metabolic diseases. The role of flavonoids in preventing the development of these and many other dangerous diseases lies in their immunostimulating (Sforcin, 2007) and antioxidant effects[15], that is, in the binding and neutralization of reactive oxygen and nitrogen species, as well as in inhibiting the synthesis of a number of other molecules involved in development of inflammatory processes (Pan et al., 2010).

The group of propolis polyphenolic compounds is not limited to flavonoids alone. The aforementioned results of a study by Japanese scientists (Kumazawa et al., 2004) show that the total content of phenolic compounds in the studied samples from 14 countries was 223±58 mg/g EEP and flavonoids accounted for 56±15% of this amount. In our aforementioned experiments with propolis from three countries, this proportion was approximately the same and amounted to 66±19%.

This raises the question of the remaining propolis phenols, which includes various derivatives of benzoic acid, aromatic alcohols, and carbonyl compounds:

---

15 *In one of the literary sources, I came across a statement that surprised me very much; allegedly "unsaturated 10-hydroxy-2-decenoic acid (10-HDA) is always found in propolis, which enters it with the secretions of the mandibular glands of worker bees" and that it determines the antioxidant properties of propolis. Once again, having carefully examined about 100 chromatograms of propolis extracts from different regions and countries, I did not find even traces of this compound. Conversely, the chemical structure of 10-HAD does not provide grounds for claiming its strong antioxidant properties. In any case, they are much weaker than those of flavonoids and other polyphenols, which will be discussed below.*

4-Hydroxybenzoic acid	4-Hydroxybenzaldehyde
Salicylic acid	2-Hydroxyacetophenone
Protocatechuic acid	Protocatechuic aldehyde
Vanillic acid	Vanillin
Isovanillic acid	Vanillyl alcohol
Vanilpropionic acid	3-Vannylpropanol
Gallic acid	3,5-Dimethoxy-4-hydroxybenzaldehyde
4-Hydroxyphenylethanol (tyrosol)	3-(4'-Hydroxyphenyl)-1-propanol

To this list should be added "simple" phenols, hydroquinone and pyrocatechol. Aromatic acids in propolis also include benzoic and anisic acids, which are not phenolic compounds but have high biological activity.

Derivatives of unsaturated cinnamic (phenylpropenoic) acid are much more widely represented in European propolis, including 1 - *p*-hydroxycinnamic (*p*-coumaric), 2 - ferulic, 3 - isoferulic, and 4 - caffeic acids.

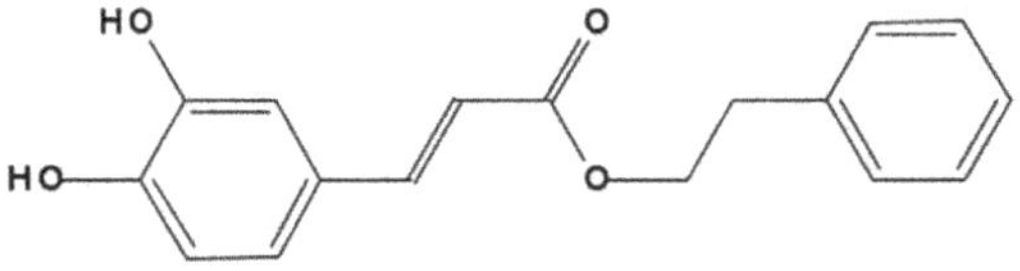

1

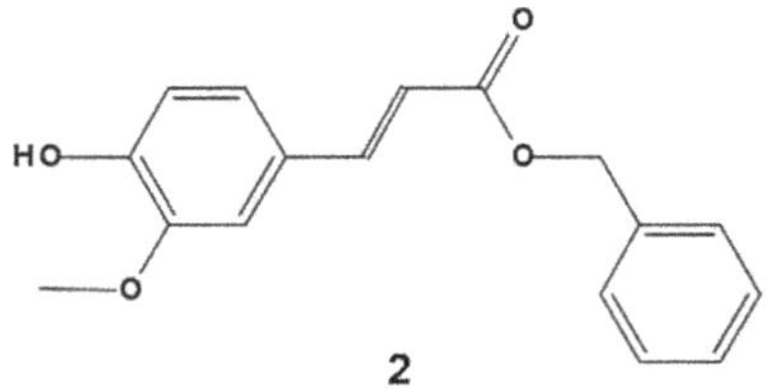

2

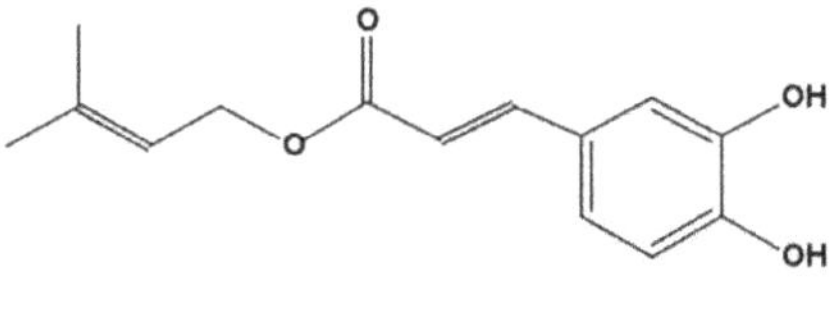

3

Figure 5.14. Hydroxylated (E)-cinnamic acid derivatives in European propolis: (1) – 2-phenylethyl caffeate (CAPE), (2) – benzyl ferulate, (3) – prenyl caffeate.

Due to the presence of a double bond in the side chain, each of these acids can exist in two isomeric (*E*)- and (*Z*)-forms. The conversion of one isomer to another easily occurs under the influence of UV radiation.

This is the basis of the protective effect of ferulic acid, which is part of many tanning creams. The acidic (carboxyl, –COOH) group is easily esterified and, therefore, many cinnamic acid esters are found in plant tissues. Below are two examples:

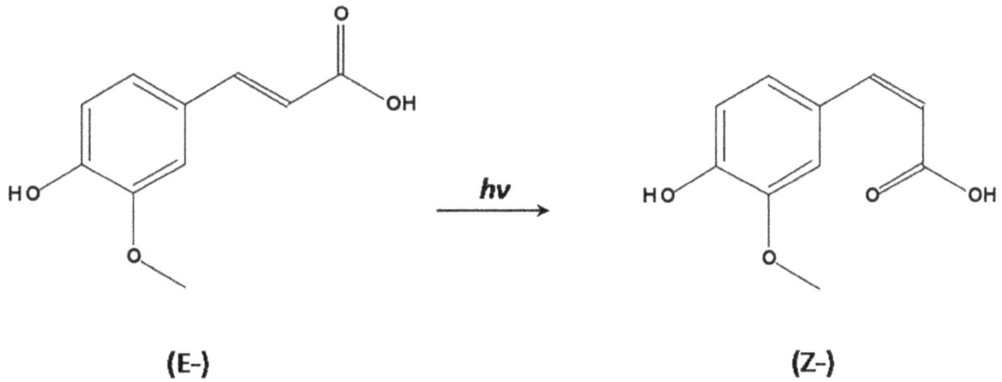

Figure 5.15. Benzyl ferulate photoisomerization

In total, in the propolis we analyzed, 95 such derivatives, also called phenylpropenoids, were identified. Additionally, another 14 related compounds, derivatives of dihydrocinnamic acid, as well as alcohols and aldehydes C6–C3, such as coniferyl aldehyde, coniferyl, and *p*-hydroxycinnamic alcohols and their esters were identified. Thus, the number of cinnamic acid derivatives in propolis significantly exceeds the number of flavonoids found in samples to date. Like other phenolic compounds, they are bactericidal, antioxidant, and anti-inflammatory. Here, we will not consider all aspects of the biological activity of cinnamic acid derivatives and refer the reader interested in this subject to a review article (Burdock, 1998), which has an extensive bibliography. Notably, caffeic acid phenylethyl ester (CAPE) has a high cytotoxicity[16] against cancer cells but not against healthy cells, which is extremely valuable (Frenkel et al., 1993; Rao et al., 1993; Beltrán-Ramirez et al., 2008). This ester is present only in the exudates of black poplars and, accordingly, in poplar and mixed propolis.

---

16 *Cytotoxicity is the ability of chemicals to induce changes in cells that lead to their death. Cancer chemotherapy is based on the use of cytotoxic drugs. Unfortunately, such drugs have a detrimental effect not only on sick cells but also on healthy cells.*

In medical practice, alcoholic extracts of propolis (EEP) and ointments are most often used. There are reports of successful application in the treatment of water extracts as well. However, the solubility of propolis in water is very low, raising the question of what the active principle is in such extracts. An alcoholic extract of propolis added to water causes an instantaneous formation of turbidity. I was firmly convinced that it was the components of wax partially dissolved in alcohol that would precipitate, but I decided to collect this matter and analyze it. It turned out that it does indeed include ethyl esters of an even number, C_{12}–C_{24} acids, typical for beeswax but their content was low (~11%). The main compounds (30%) were flavonoids, sakuranetin, isosakuranetin, 4-methoxy sakuranetin, pinocembrin, galangin, and esters of cinnamic acids, mainly *p*-coumarates (31%). Another 17% were sesquiterpenoids. All these substances dissolve well in alcohol but very weakly in water. Therefore, when mixing an alcoholic extract in water, they form a turbid suspension, which does not affect their absorption in the body. It is unclear how water extraction from propolis is useful. It was not possible to find an answer to this question in the literature; thus, we decided to carry out a study using the available water extraction methods of two propolis samples of different origins.

Stirring finely ground propolis in hot water for four hours gave a slightly yellow solution. Extraction of the aqueous extract with ethyl acetate and subsequent analysis showed that it contained benzoic acid as the main component, as well as unsubstituted cinnamic acids. Flavonoids were the main constituents of EET in both propolis samples but their content was low (0.2–2.3%)in aqueous extracts. It turns out that the aqueous extract does indeed contain biologically active compounds and their composition depends on the plant precursor. This can be seen from the Table 5.4.

Table 5.4. Comparison of the composition of polar compounds in aqueous propolis extracts of two different types

Compound	Relative composition (%)	
	Birch-type propolis (Russia)	Poplar-aspen propolis (Poland)
Hydroquinone	0.4	0.5
Benzoic acid	52.0	15.2
4-Hydroxybenzoic acid	1.4	2.0
Vanillic aacid	trace	0.3
Gentisic acid	-	0.5
(E)- & (Z)-p-Coumaric acids	22.3	29.5
Isoferulic acid	-	5.2
(E)- & (Z)-Ferulic acids	6.4	5.5
Caffeic acid	-	8.0
Sakuranetin	0.2	-
Pinobanksin	-	2.3
Glycerol phenylpropenoids	-	1.0
Valillin	4.7	4.6
Aromatic aldehydes	1.3	1.9
Sesquiterpene alcohols	1.3	0.2

Notably, even in the extract from poplar propolis, there are no esters of cinnamic acids that characteristic of it. Some of them cause an allergic reaction (Hausen et al., 1987a, 1987b; Khlgatyan et al., 2008); this applies to 3-methyl-2-butenyl caffeate, LB-1 factor.

Therefore, poplar propolis in the form of an alcoholic extract cannot be used by people with hypersensitivity to allergens. However, the aqueous extract can be used by them. It will not hurt children, as it is contained within common fruits and many vegetables. These unpleasant esters are completely absent in northern propolis, originating from the exudates of aspen and birch buds. Propolis of this type also contains esters of cinnamic acids, but with different physicochemical and biological properties. These are the phenylpropenoids of glycerol and phenylpropenoids of sesquiterpene alcohols that were previously mentioned.

Esters of glycerol are well known to all as vegetable oils and animal fats are composed of them. In oils and fats, glycerol is associated with fatty acids, mainly with a long chain of 16 or 18 carbon atoms, such as palmitic, stearic, oleic, and linolenic, among others. It is only relatively recently that other unusual glycerides have been discovered in nature. In the molecule of which, instead of fatty acids, there are residues of other organic acids, including cinnamic acids. Phenylpropenoid glycerides were first discovered by Popravko et al. (1982) in aspen buds and Russian propolis. They managed to isolate and identify two compounds of this class. Later, several such glycerides were found in Bulgarian, Swiss, and Turkish propolis, originating from the areas of aspen and *Populus euphratica* (Bankova et al., 2002; Popova et al., 2005). Our studies revealed 39 such glycerides in aspen bud exudates (Isidorov et al., 2008), most of which were also found in propolis. These compounds deserve a closer look to characterize their properties. The structural formula of two of the phenylpropenoid glycerides of aspen buds are shown in Fig. 5.16.

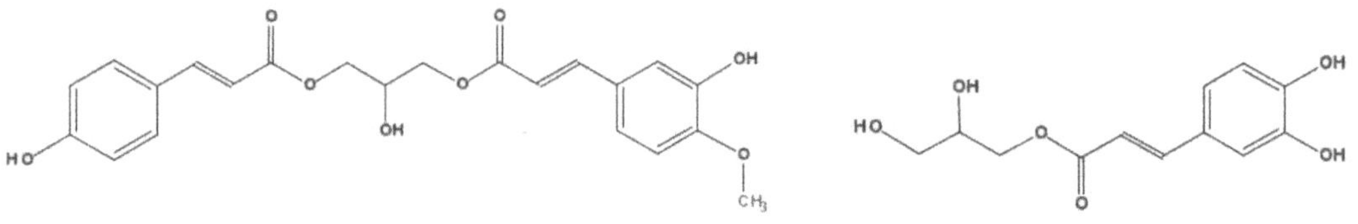

1 2

Figure 5.16. Chemical formula of 1-p-coumaroyl-3-feruloyl glycerol (1) and 1-caffeoyl glycerol (2)

The glycerol molecule can be combined with residues of one or two, identical or different, cinnamic acids. Compounds with three such fragments have not been found in nature, but there are compounds in which the third –OH group of glycerol is acylated, that is, replaced by an acetic acid residue. The physicochemical properties of various phenylpropenoids depend on the number and nature of the substituents. The most polar and water-soluble compounds are compounds with only one substituent (monoglycerides). Di-substituted glycerides are less polar and tri-substituted glycerides are even less polar. With the same degree of substitution, the most polar glycerides, those containing residues of caffeic acid, are two phenolic hydroxyl, as shown in Fig 5.16 (formula 2).

This means that phenylpropenoid glycerides of aspen buds and propolis can exhibit a wide range of hydrophilic-hydrophobic interactions with cell membranes and lipophilic and hydrophilic fragments of various biomolecules. The presence of phenolic groups in the residues of cinnamic acids determines the antioxidant and bactericidal properties of such glycerides. This makes phenylpropenoid glycerides a very interesting object of biomedical research. Unfortunately, to date, not many such studies have been carried out, although their results are promising; some of these glycerides have been found to have anticancer and antiproliferative activity (Gunasekera et al., 1981; Banskota et al., 2000). It has been established that they are not degraded by intestinal lipases and, therefore, can directly enter the bloodstream. It can also be argued with a high degree of probability that phenylpropenoid glycerides are safe since many of them are found in corn and wheat grains.

Phenylpropenoids of terpene alcohols are another group of practically unexplored biologically active components of propolis. The first representatives of this series, the phenylpropenoids of monoterpenol geraniol ($C_{10}H_{18}O$), geranyl *p*-coumarate, and geranyl (*E*)-caffeate, were reported by English researchers (Greenaway & Whatley, 1990; 1991). The first of them was identified in the exudate of the buds of the American balsamic poplar (*P. balsamifera*) and the second, in the exudate of the buds of the Himalayan poplar (*P. ciliata*), as well as in the composition of propolis collected in England. Hence, it can be concluded that not only the Himalayan poplar contains this compound.

More complex phenylpropenoids of sesquiterpenols ($C_{15}H_{24}O$) were first isolated from the extracts of silver birch buds (*B. pendula* Roth; synonym *B. verrucosa*) by Russian chemists (Vedernikov et al., 2007). These were *p*-coumarate of 6-hydroxy-β-caryophyllene, 14-hydroxy-β-caryophyllene, 9-*epi*-14-hydroxy-β-caryophyllene, and 14-hydroxy-α-humulene. We managed to expand the range of these unusual compounds due to ferulates, esters of the same sesquiterpenols and ferulic acid.

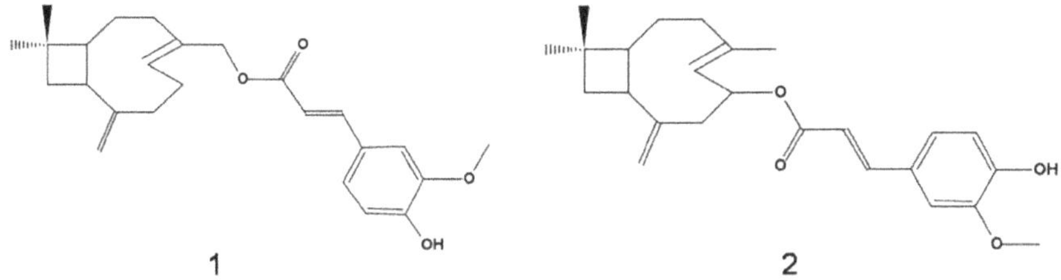

1

2

Figure 5.17.Chemical formula of 14-hydroxy-β-caryophyllene ferulate and 6- hydroxy-β-caryophyllene ferulate.

Interestingly, we found these phenylpropenoids in exudates and extracts of buds of downy birch (*B. pubescens*) but not silver birch. In the buds of the latter species, collected in Russia, Latvia, and Poland, we did not find even a hint of them. To resolve this contradiction, we contacted a leading specialist in the field of birch biology, Lidia Vetchinnikova, who works at the Forest Institute of the Karelian Branch of the Russian Academy of Sciences. She sent us buds collected from birches grown at the experimental station of the Institute, for which the taxonomic affiliation has been reliably established. Moreover, we obtained the buds of not only these two species but also several other forms of silver birch, Karelian birch (*B. pendula* var. *carelica* Mercklin), dalecarlic birch (*B. pendula* var. *dalecarlica* Schneid.), and its varieties (*B. pendula* .var. *dalecarlica* Crispa). In addition to this, we also obtained buds of silver and downy birches from the arboretum of the Institute of Dendrology of the Polish Academy of Sciences. Analysis of exudates and extracts of these samples unambiguously confirmed our results; the buds of silver birch do not contain phenylpropenoids, nor do they contain flavonoids, which the same Russian chemists reported in another publication (Galashkina et al., 2004). One of two things occurred, either the authors of these works, impeccable from the point of view of chemistry, made a mistake in the botanical identity of the trees they studied, or they dealt with hybrids of silver and downy birch as these species, although they have a different set of chromosomes, sometimes interbreed with each other. I am more inclined to the latter assumption. In their next publication (Vedernikov & Roshchin, 2010), the authors reported the discovery of compounds characteristic of both species, sesquiterpenoids of downy birch and triterpenoids (dammaradien-3-one and dipterocarpol) of silver birch. This is not surprising as hybrid forms often have biosynthetic and metabolic pathways characteristic of both parents (Isidorov et al., 2019).

Unfortunately, chemists are not always careful and cautious in determining the taxonomic affiliation of the plants in question. For example, Turkish authors (Silici & Kutluca, 2005) reported on the analysis of local propolis allegedly collected by bees from the buds of *Populus tremuloides*, which is an exclusively North American species similar to European aspen (*Populus tremula*).

The phenylpropenoids of sesquiterpenols are phenolic compounds and, therefore, must have antioxidant and bactericidal properties characteristic of all phenols. Not so long ago, Japanese scientists (Uwai et al., 2008) studied the biological activity of a number of phenylpropenoids, including geranyl-(*E*)-caffeate, identified in

English propolis, and farnesyl-(*E*)-caffeate, specially synthesized by the authors. It turned out that both of these compounds actively inhibit the synthesis of nitric oxide (NO), which plays an important role in the development of inflammatory processes. Like flavonoids, they exhibit anti-inflammatory properties. It can be assumed that other phenylpropenoids discovered by us in birch-type propolis have the same attractive properties from the point of view of the therapeutic effect[17]. However, these compounds, as well as numerous phenylpropenoid glycerides of aspen-type propolis, have not been well studied.

There are few publications that would provide data on the direct effect of volatile substances of propolis on microorganisms. There is no doubt, however, that they have antimicrobial properties. A significant fraction of propolis VOCs are formed by lower alcohols and acids, as well as terpenoids. The antiseptic properties of alcohols and acetic and other lower acids are well known. Experts are also aware of the antimicrobial effect of plant essential oils, which are mainly composed of terpenoids. For example, in microbiological tests, it was shown that sesquiterpene alcohols and their acetates of the caryophyllene series (found in birch-type propolis) are active against microorganisms, such as *Escherichia coli*, *Staphylococcus aureus*, *Micrococcus luteus*, *Pseudomonas aeruginosa*, and *Candida glabrata* (Demirci et al., 2000). Greek researchers (Melliou et al., 2007) demonstrated the antimicrobial activity of volatile fractions of local propolis against these species, as well as against *S. epidermides*, *Enterobacter cloacae*, *Klebsiella pneumoniae*, *Candida albicans*, and *C. tropicalis* (the latter two, as well as *C. glabrata* are fungi that are pathogenic to humans).

In conclusion, I will share one more observation that surprised me. It turns out that dry propolis can retain its properties for a very long time. I was convinced of this by preparing and analyzing the extract of propolis, which had been stored in my friend's family for over 20 years. It was possible to split this lump with great difficulty as it had hardened so strongly over the years but according to the results of the analysis, I could not distinguish it from the one collected in the same area this year. In it, "signals" from birch and aspen with a small admixture of poplar bud exudates were clearly seen.

---

17 *In the exudates of the buds of the Litvinov birch (B. litwinowii) that grows in the Caucasus and Turkey, we were able to detect 12 phenylpropenoids of sesquiterpenols ($C_{15}H_{24}O$ and $C_{15}H_{26}O$) (Isidorov et al., 2015). It can be expected that in propolis from the distribution area of this birch, these phenylpropenoids will also be discovered over time.*

Chapter 6 What's new about honey?

6.1. Is it worth continuing to research it?

What new can be said about a product that has received more than 13,000 publications in the journals of only two scientific publishers over the past ten years? After all, not 10 years ago, but much earlier, people began to write about honey and not only in the journals of the Elsevier and Springer publishing houses. As early as 1886, G.A. Koschevnikov, the future discoverer of the gland named after him, was touched by the huge number of books and scientific articles about bees and their products.

Thus, it can be assumed that for 120–150 years around the world in the journals of chemical, biological, medical and technological profiles, about 100,000 scientific papers were published, and many of them are devoted to the study of honey. This circumstance could cause sadness to young scientists determined to devote themselves to the topic of honey. However, I can say with all responsibility that there is no reason to be pessimistic. First of all, because of the development in chemical analysis methods and the progress in other branches of science and technology, new possibilities for studying the origin, chemical composition, and properties of honey open up. In addition, the very name honey is collective and there is probably no daredevil who will declare his readiness to list all its varieties. During all-Russian honey trade exhibition in St. Petersburg in 2011, where professional beekeepers from all over Russia came, I counted over 60 such varieties, each with their own secret and many more that are undiscovered. For example, the relationship between honey and its plant precursor, flower nectar, and honeydew has not been fully explored. Much is also unclear about the biologically active substances or their complexes, responsible for the antimicrobial and healing properties of honey.

In this chapter, the reader will not find the characteristics of different varieties of honey, nor the provisions on their use in various diseases, nor disease prevention. All these issues are very well exposed in many publications, of which I will mention only a few. In this chapter, I will try to address issues related to the origin of honey, the specific features of its composition, and some properties that are much less discussed in the generally available literature.

6.2. Sweet contribution of plants paid to bees

A strange and magical power
The flower is amazingly sparkling
In it, combined with heavenly consolation
There are sinful charms of the earth
E. Varzhenevskaya

The history of bee honey, as is well known, begins with the collection of its plant precursors, flower nectar or honeydew. It is surprising that the chemical composition of these raw materials has, so far, been poorly studied, with the exception of the sugar content. The most plausible explanation is that it is related to the difficulty of obtaining enough nectar for detailed chemical analyses. After all, with rare exceptions, there are only a few milligrams of nectar in a single flower of a honey plant and obtaining it is not an easy task. They are mainly made with the use of special glass micropipettes presented in the photographs below, which, as a rule, should be produced by the experimenter (Jabłoński, 2003).

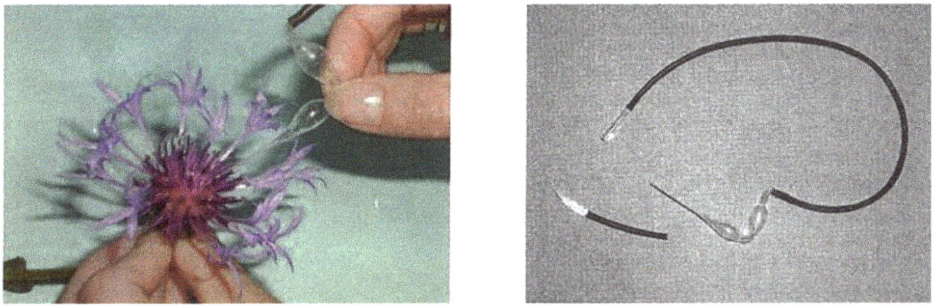

Figure 6.1. Taking cornflower nectar samples and a set of tools to do this (Photo: Z. Kołtowski)

The interest in sugars in nectar is understandable because this is a practical issue; beekeepers are primarily interested in the size of the nectar flow. However, the taste properties and unique aroma, as well as the healing properties of the future honey depend on other characteristics of nectar, such as the content of secondary metabolites. These include various volatile substances, fatty and aromatic acids, lipids, amino acids, vitamins, and even psychoactive alkaloids, such as caffeine, nicotine, and amygdalin (Baker, 1977). Our knowledge of these nectar components is fragmentary. My own experience in this field is also very modest (Isidorov et al., 2011a) and, therefore, mainly the results of a few studies by other authors will be presented here.

The sugars produced during photosynthesis, called primary metabolites, are a source of carbon and energy necessary for plant development at all stages. Photosynthesis proceeds through many successive stages, in which part of the solar radiation energy absorbed by chlorophyll molecules is lost and converted into heat. For this reason, even in the case of excellent agricultural plant species, the photosynthesis efficiency factor, even in optimal conditions, does not exceed 6–7%. Secondary metabolites are synthesized with much lower efficiency. All of this means that plants have to invest a great deal of energy in producing the nectar components.

Unfortunately, the energy losses of these processes were estimated only for single plants. For example, it was explained that in American milk thistle (*Asclepias syriaca*), about 37% of sugars synthesized during the day are converted to nectar. Therefore, the honey yield of this plant is estimated at about 600 kg/ha (Lipiński, 2010). In the case of alfalfa (*Medicago sativa*), the energy invested in nectar production is twice that needed for seed production (Pacini et al., 2003). These calculations are without considering the additional energy losses for transport and nectar separation. In addition, flowers must maintain the concentration of sugars in the nectar at a certain level and compensate for the evaporation of water. Furthermore, nectar is an expensive "tribute", which is why many plant species have developed energy-saving strategies based on nectar resorption (Nepi & Stpiczyńska, 2008). The phenomenon of reverse nectar absorption was first discovered in 1878 (Bonnier). It has now been explained that resorption occurs faster in pollinated flowers than in non-pollinated flowers. Night resorption is very economical because it protects sugars from being stolen by foreign insects that do not participate in cross-pollination.

Notably, in the course of evolution, flowering plants adapted to the requirements of pollinators and this applies not only to the sugar content in the nectar. For example, Baker (1977) explained that the nectar of flowers pollinated by day butterflies and many wasp species that do not use plant pollen is high in amino acids and peptides. However, it is much lower in the nectar of flowers visited by bees.

The same author reported on another interesting phenomenon. Some species of oriole in Mexico protect the territory where erythrine (Erythrina breviflora) grows. The nectar of its flowers is exceptionally rich in amino acids (on average 3.9 mg/ml) and contains all essential amino acids. Birds pollinate flowers and get food from them that is rich in nitrogen and, therefore, they do not have to hunt insects in search of proteins. The phenomenon of pollination of flowers by birds is called ornithogamy.

The adaptation of flowering plants to pollinators is manifested in the content of sugars and amino acids in the nectar. It also contains substances capable of suppressing the appetite of uncomfortable insects, such substances are called food detectors or antifeedants. Apart from the aforementioned alkaloids, they include some phenolic compounds and some non-proteinogenic amino acids and do not harm their pollinators (Singaravelan et al., 2005). For example, the caffeine in the nectar of orange blossoms does not harm bees in any way but it does have a strong effect on the central nervous system of insects that are eager to enjoy it for free. It was noticed that the presence of antifeedants in the nectar of flowers of various species is directly proportional to the species richness of insects in a given region. For example, in the insect-infested subtropical zone, the nectar of more than 50% of the studied plant species contained phenolic substances, while in plants from alpine meadows, where the diversity of insects is significantly lower, phenols are found 2.5 times less frequently. In the first case, alkaloids were found in the nectar of 12% of plant species and in the second, they were absent (Baker, 1977).

Flowers attract bees with their colour but also with their scent. The smell of nectar is given by the volatile substances contained in it, which, as shown in Fig. 6.2, may be very numerous (nectar samples of these plants were collected in 2011 by Dr. Zbigniew Kołtowski).

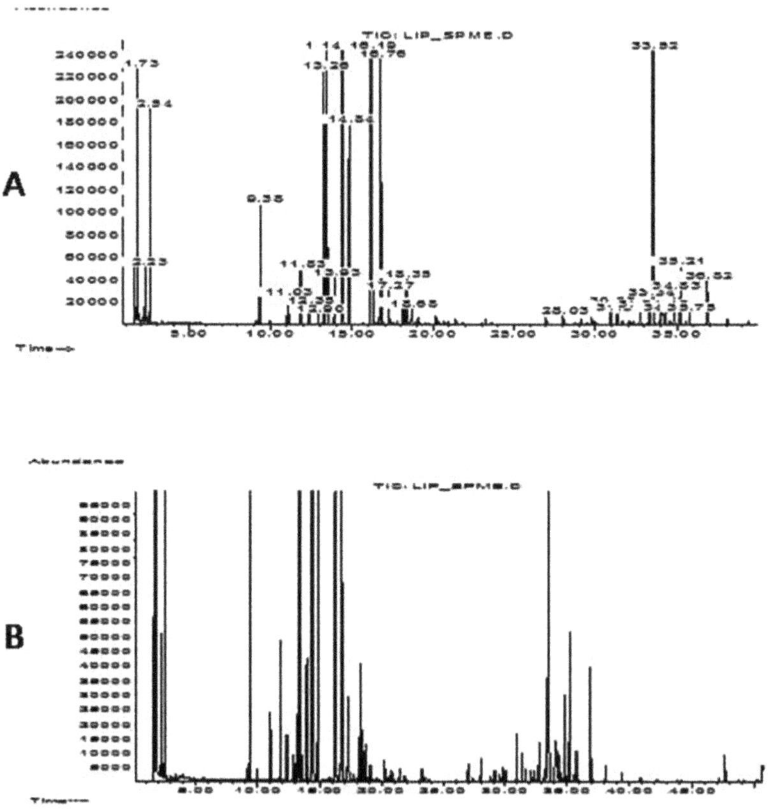

Figure 6.2. Chromatograms of volatile secretions (VOC) of the nectars of blackcurrant flowers (a) and broadleaf linden (b).

I have the impression that the smell of nectar and flowers are not the same. This thought comes from comparing the composition of volatile flower exudates and nectars from rapeseed and linden; in the case of whole flowers, the VOC spectrum was richer. Indeed, volatile substances can give off not only nectar, but also flakes, stamens, birthmarks, and pollen. It is possible that these secretions may be the main scent stimuli for bees. However, I have no basis to draw any specific conclusions on this subject. Therefore, more detailed research is necessary.

The second plant precursor to honey is honeydew, the cell sap sucked out of plant tissues by parasitic insects. It is also a tribute but no longer voluntary because the plants receive nothing in return. The main producers of honeydew in the mid-latitude climatic zone are insects of the order *Homoptera* suborder and the aphid (*Aphidina*), numbering tens of thousands of species, as well as the scale insects (Gałuszka et al., 1996). The most common species of aphids belong to the families

Lachnidae, *Phillaphididae*, *Chaitophoridae*, and *Thelaxidae*. There are about 600 species of these insects in Poland. Out of about 6,000 known species of scale insects (*Coccina* insects) in the temperate zone of Europe, about 150 have been described, of which insects from the *Coccidae* family are the main honeydew producers.

Figure 6.3. Aphids feeding on pine trees. (Photo: V. Vinogorova)

Honeydew is created in the following process. At the beginning the insect pierces the tissue covering the plants with its dagger-like mouthpiece. It penetrates into the phloem, the tissue carrying organic substances synthesized by the plant, and the cell sap, under high pressure, enters the digestive tract. This juice contains a lot of sugars, the content of which significantly exceeds the energy needs of the insect but not enough essential amino acids and peptides for the parasite. These nitrogenous substances are captured by the body of the insect from the juice and the undigested part with a high sugar content goes to the filtration chamber and through the intestine are thrown outside as drops. When large colonies of aphids or scarlet colonies appear on one leaf of the plant, the number of drops of juice they emit becomes so large that they fall on the lower leaves and cover them with a shiny layer. This is where the Polish name, spadź (fall off), of these secretions comes from.

According to Gałuszka et al. (1996), on deciduous trees, honeydew most often occurs at the end of May and the beginning of June (on lindens during flowering). In the case of conifers, the peak of insect feeding falls in the second half of July and at the beginning of August. However, these dates may change from year to year, depending on weather conditions. Honeydew appears not only on trees, but also on herbaceous plants. Infrequently, the fields of grain untreated with insecticides are attacked by aphids and cause abundant honeydew.

Studies of the chemical composition of honeydew were begun in the 1920s by a Russian chemist, Professor I.A. Kablukov. He explained that 85–95% of honeydew dry matter is made up of sugars. Later studies have shown that it also contains sugars that are not present in the sap of plants on which insects feed. This means that these substances are produced in the body of parasites. Of these secondary sugars, the trisaccharide, melezitose, is the most abundant. The production of melezitose is of great importance to the physiology of sucking aphid-like insects. They help to prevent osmosis stress. However, for bees, its high content is harmful. Bee enzymes easily break down melezitose into glucose and the disaccharide, turanose. The latter is an isomer of sucrose but it is poorly absorbed by bees. Therefore, in winter conditions, bees that eat honeydew accumulate a lot of faeces. Moreover, melezitose crystallizes easily. Honeydew is literally cemented in the combs during the winter and lack of liquid water, causes the bees to starve. A lot of melezitose (up to 50–60% of the total sugar content) is produced by the *Cinara laricis* aphid (speckled larch *aphid*), which feeds on larch (Gałuszka et al., 1996). Beekeepers sometimes call honey obtained from such honeydew, cement.

The high content of mineral salts in honeydew, which is 8–10 times more than in nectar honey, also affects bees disastrously. An excess of salt, especially salt containing sodium, interferes with the secretion of water from the faeces. As a result, the posterior intestine becomes overfilled with watery faeces, causing diarrhoea and death of the bees. Therefore, experienced beekeepers never leave it for winter food. However, melezitose is not harmful to humans and the high content of mineral substances in honeydew honey increases its consumption value. Therefore, it enjoys consistently high demand.

Unfortunately, I was unable to collect and analyse this substance from coniferous plants. I had more success with some deciduous trees, including English oak (*Quercus robur*), linden (*Tilia cordata*), aspen (*Populus tremula*), and mirabelle plum (*Prunus domestica* L. ssp. *Syriaca*). Honeydew from the leaves of these trees contained relatively small amounts of melezitose (0.5–0.6%) and only in linden, its content increased to 11%.

Other trisaccharides in the honeydew of all four trees are represented by raffinose (0.1–0.8%), maltotriose (0.1–1.2%), 1-kestose (0.6–3.3%), and erlose, in which the most honeydew were obtained from lime (13%) and oak (30%). Honeydew of mirabelle plum was distinguished by a very high content of isomeric sugar alcohols, mannitol (6.3%) and sorbitol (45.6%). The richest in sugars easily assimilated by bees (60–66% of fructose, glucose, and sucrose) were lime and oak leaf washes. The presence of quinic acid and cyclic alcohol, quercitol, were characteristic of oak honeydew.

Harvesting the honeydew directly from aphids or scale insects (*Coccoidea*) is even more difficult than nectar from flowers. For this reason, practically nothing is known about the content of its volatile substances and other secondary metabolites. There is no doubt that they exist as the aroma of honeydew honey proves it.

6.3. The way from nectar to honey – the order of magical transformations

The species name *Apis mellifera* (which translates from Latin as a *bee that brings honey*) first appeared in 1758 in the tenth edition of Carl Linnaeus' fundamental work Systema Naturae. However, already in the twelfth edition (1766), Linnaeus proposed a different name, *Apis mellifica*. This name translates as a *honey producing bee*. The new name, however, did not catch on, although it is mentioned in the publications of some authors (for example, by M. Barbier, 1981)[18]. Linnaeus admitted his mistake and tried to correct it. It follows that he concluded that the bee brought nectar to its nest and only there was it transformed into honey.

The detection of the enzymatic mechanism of decomposition of sucrose into simple sugars, as well as some observations and experiments in recent years, prove that the bee brings a substance to the hive that is already strongly different from nectar, not only in terms of the content of various forms of sugar. Swiss scientists (Naef et al., 2004) performed an interesting experiment where they examined the composition of the nectar extracts of small-leaved linden (*Tilia cordata*) flowers, the content of honey sacs from bees returning from the collection of this nectar and captured on the hive's arrival board, and mature linden honey from the same family. The composition of the nectar turned out to be complicated; it contained many terpene substances, aromatic alcohols, and aldehydes, as well as small amounts of alkaloids (caffeine, theophylline, and traces of nicotine). In the honey sac of bees, not only were these substances found but also many new ones, which, according to the authors, were formed as a result of the breakdown of more complex components by

[18] *"Apis mellifica" is also the name of a homeopathic medicinal product without therapeutic indications (Boiron S.A., France).*

enzymes from the glycosidase group contained in bees' saliva. Linden honey extract turned out to be even more complicated with more new compounds appearing in it, including mentofurans (products of terpenol cyclization) and terpene acids (products of terpenol oxidation). These changes are clearly visible in the chromatograms presented in Fig. 6.4.

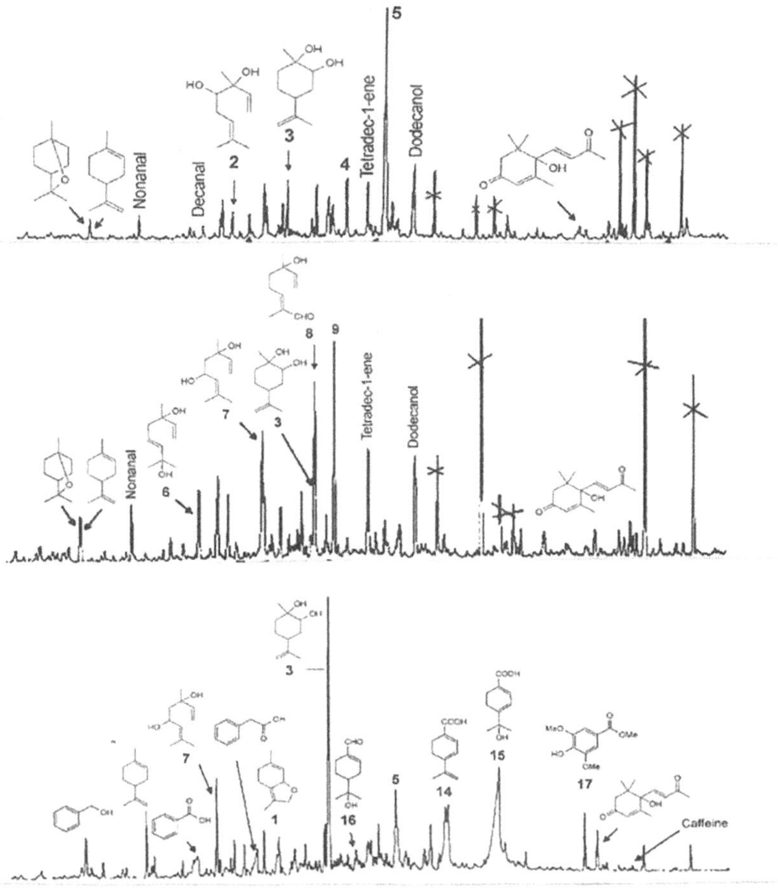

Figure 6.4. Chromatograms of extracts (top to bottom) of linden blossom nectar, honey sac content of bees, and linden honey. The peaks of environmental pollutants are deleted (Naef et al., 2004).

The results of this interesting experiment, however, do not conclude whether the changes in the composition of nectar occur chemically or under the influence of some undetected enzymes.

Greek authors are slightly more explicit on this topic (Alissandrakis et al., 2010). They fed bees with sugar syrup with the addition of linalool, the main terpene substance of citrus honey, and observed the products of its transformation. Analysis of the composition of the obtained "pseudo-honey" was carried out on the 8th, 13th, and 21st days after feeding the bees. As a result, it was explained that the products of its cyclization and oxidation were formed from linalool. Their content increased over time, with highest increase between the 8th and 13th day of ripening of «pseudo-honey». In their opinion, the most reliable changes take place with the participation of the enzyme P-450 hydroxylase, produced in the bees. This assumption may be correct but has not been sufficiently substantiated, as the authors rely on analogies in plant biochemistry.

The pioneering work of Swiss scientists undoubtedly proves that (bio)chemical changes of nectar begin from the moment it is absorbed into the honey sac of the bee. There is a lot to explain about the mechanisms of these changes. What else is known about the role of bees in changing the chemical composition of nectar and what components of honey are of animal origin?

The latter includes enzymes, such as invertase (β-fructofuranosidase), α- and β-amylase, and glucose oxidase. Under the action of invertase, sucrose is broken down into glucose and fructose. Diastatic enzymes (α- and β-amylase) break down polysaccharides. For example, β-amylase hydrolyzes amylose, one of the major polysaccharides of starch, forms the disaccharide maltose. The bactericidal properties of honey are significantly related to glucose oxidase because its action on glucose produces a strong aseptic, hydrogen peroxide (H_2O_2):

$$C_6H_{12}O_6 + O_2 + H_2O \rightarrow C_6H_{12}O_7 + H_2O_2$$

This reaction was used to manufacture an oxygen absorbing device inside packages with oxygen-sensitive food products, including peanuts and powdered milk, among others. The antioxidant package is a gas-permeable foil bag containing a granulated mixture of moist glucose and glucose oxidase. As a result of the reaction, oxygen is slowly consumed and at the same time, significant amounts of hydrogen peroxide are formed, which in higher concentrations can inactivate glucose oxidase. To prevent this from happening, a small amount of another enzyme, catalase, is added to the absorber to break down the peroxide formed. In this case, the summary reaction equation is: $2\ C_6H_{12}O_6 + O_2 \rightarrow 2C_6H_{12}O_7$

Glucose oxidase is distinguished by very high selectivity; of all 64 cyclic forms, 16 monosaccharides (aldohexoses) catalyze the oxidation of only β-glucopyranose. Gluconic acid, $C_6H_{12}O_7$, is also a product of glucose oxidation. Of all the acids contained in honey, this one has the largest share.

The trehalose disaccharide, which was discussed in the third chapter, also has animal origin as one of the main sugars in the bee's hemolymph. Synthesized in the cells of the fat body from two glucose molecules, trehalose is transported by the hemolymph to tissues that need glucose (Blatt & Roces, 2001). There it again breaks down into two glucose molecules with the participation of the enzyme trehalase. This means that trehalose is nothing more than a transport form of glucose.

This disaccharide, uncharacteristic for plants in this climate zone, is produced by the action of another bee enzyme, maltase. It has the ability to attach part of sucrose, either glucose or fructose, to another saccharide molecule. In the case of bees, this causes the formation of not only trehalose disaccharide but also new trisaccharides, such as fructomaltose and maltotriose; the amount of which is small in honey.

Another group of honey ingredients that are absolutely unusual for the plant world are royal jelly hydroxy acids, which were already mentioned in the third chapter of the book. The fact that they are not random admixtures is indicated by the fact that they were found in over 150 samples of honey (Isidorov et al., 2011b). These included rapeseed, buckwheat, lime, and honeydew honeys from Poland, heather honeys from Poland and Scotland, acacia honeys from Poland and Austria, hawthorn honeys from Russia, orange and eucalyptus honeys from Morocco, and even exotic nectar honey from New Zealand manuka flowers. In all of these samples in different amounts and combinations, 8 such acids were identified, including 7-

and 8-hydroxyoctanoic, 3-, 9- and 10-hydroxydecanoic, 3,10-dihydroxydecanoic, unsaturated hydroxy acids: 9-hydroxy-2- decene (9-HDA), and 10-hydroxy-2-decene (10-HDA), as well as unsaturated 2-octene-1,8-dioic and 2-decene-1,10-dioic dicarboxylic acids. In the highest amounts, all of these acids were found in silver fir (*Abies alba*, native to Europe) honeydew honeys (Fig. 6.5). In seven samples from southern Poland, the total content of these compounds was in the range of 24–41 µg/g and the presence of 10-HDA was 12–20% of this content (Isidorov et al., 2011b).

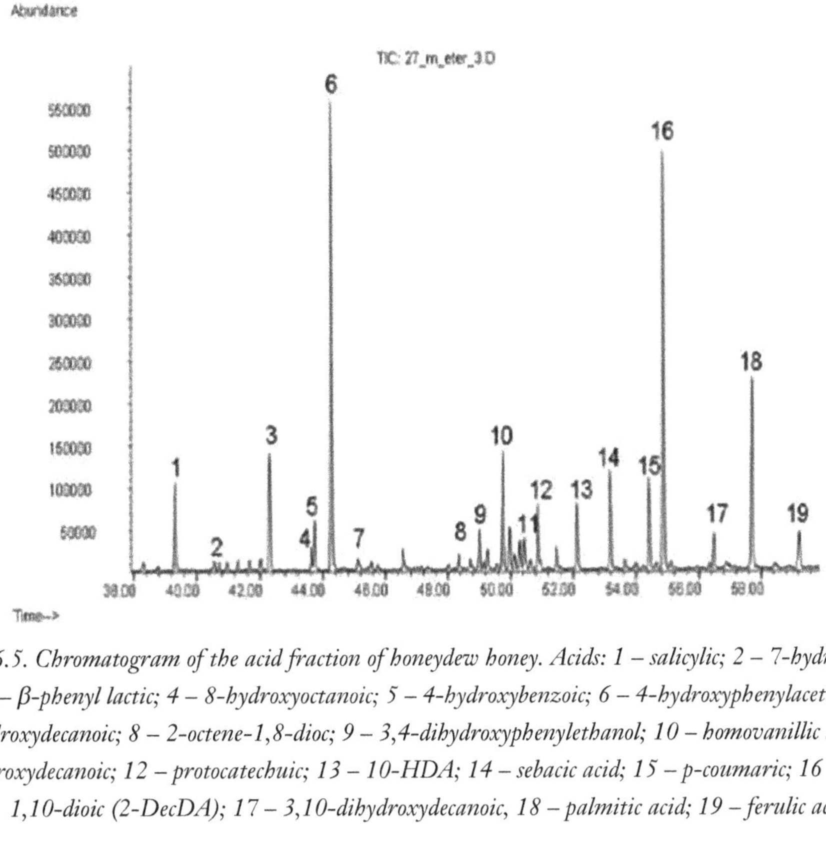

Figure 6.5. Chromatogram of the acid fraction of honeydew honey. Acids: 1 – salicylic; 2 – 7-hydroxyoctanoic; 3 – β-phenyl lactic; 4 – 8-hydroxyoctanoic; 5 – 4-hydroxybenzoic; 6 – 4-hydroxyphenylacetic; 7 – 3-hydroxydecanoic; 8 – 2-octene-1,8-dioc; 9 – 3,4-dihydroxyphenylethanol; 10 – homovanillic acid; 11 – 10-hydroxydecanoic; 12 – protocatechuic; 13 – 10-HDA; 14 – sebacic acid; 15 – p-coumaric; 16 – 2-decene-1,10-dioic (2-DecDA); 17 – 3,10-dihydroxydecanoic, 18 – palmitic acid; 19 – ferulic acid.

Previously, the quantitative content was determined only for one of these substances, 2-decene-1,10-dicarboxylic acid (2-DecDA). According to New Zealand authors (Tan et al., 1988), it was 40.3±57.2 µg/g in white clover honey and 31.1±20.7 µg/g from the nectar of manuka flowers. We detected the largest amounts (up to 100 µg/g) of 2-DecDA in Polish linden honeys. It can be assumed that this acid is formed by the enzymatic oxidation of 10-HDA (Plettner et al., 1996). This is confirmed by a fairly strong negative correlation between the concentration (C) of these two acids; for 20 samples of different varieties of honey, it can be described by the equation:

$$C^{2\text{-DecDA}} = 6.846\ C^{10\text{-HDA}} - 17.776\ (R = 0.881)$$

How and when do these components of royal jelly end up in honey? What is their influence on the properties of honey? These are the questions that deserve attention. In our opinion, they will go to the nectar processed into honey from the glands of young hive bees that receive the nectar flow from foraging bees. It is known that a bee returning with nectar or honeydew does not fold it immediately into the cell but passes it on to two or three workers aged 8–16 days. They pass these, in turn, via trophallaxis (see chapter 2.3) to other bees (Wainselboim et al., 2003). Receiving bees are repeatedly drawn into the honey sac and again release a drop of liquid at the end of the tongue. During these activities, this liquid becomes enriched with enzymes and at the same time, loses water due to evaporation. Only after this does the bee attach the thickened droplet to the wall of an empty or already partially filled cell. It can be assumed that during the circulation of the processed nectar inside the nest, it is enriched not only with the enzymes of the salivary glands but also with the secretions of the glands that produce royal jelly. The case is different during the main nectar flow. Then, the foraging bees attach the brought nectar droplet to the cell wall by themselves. In such a situation, water evaporation occurs mainly due to the constant ventilation of the socket and expulsion of moist air. It has also been noticed that worker bees bring some of the droplets from one cell to another. At the same time, they enrich the secretions of their glands.

If we talk about the effects of enriching honey with royal jelly, they are undoubtedly positive. For example, royal jelly has bactericidal properties and the main unique acid it contains is 10-HDA (Isidorov et al., 2018). Many scientists associate the antimicrobial properties of honey with the presence of hydrogen peroxide, H_2O_2 (Molan & Russell, 1988; Molan, 1992; Brudzynski et al., 2011)[19], methylglyoxal (Adams et al. 2008; Atrott & Henle, 2012; Cokcetin et al., 2016), or polyphenols (Weston et al., 1999; 2000). However, experiments show that the

19 *In fact, there is no hydrogen peroxide in mature honey. It appears only after dilution with water, when conditions are created for the activity of the enzyme glucose oxidase. Therefore, honey as an antiseptic agent must first be diluted in water and left for some time (for example, from evening to morning).*

elimination of hydrogen peroxide and methylglyoxal from honey only partially reduces the antimicrobial activity of honey (Kwakman et al., 2010a). This suggests that some other components of honey play an important role in this residual activity. The hypothesis about the influence of polyphenols has been disproved because the content is too low and cannot cause the observed activity. Therefore, the aseptic properties of honey have not been fully explained.

In this context, the experiment of the Swiss scientist Stefan Bogdanov (Bogdanov, 1997) is very interesting, in which, after the initial elimination of H_2O_2, 8 genuine honeys were divided into several fractions. He also included samples of pseudo-honey, obtained by feeding bees with pure sugar syrup. Each of these fractions was tested for residual non-peroxide activity against two bacterial species: *Staphylococcus aureus* and *Micrococcus luteus*. It turned out that the aseptic activity of individual fractions against both bacterial species decreased in the following order: acid fraction>alkaline fraction≈neutral, non-volatile fraction>volatile fraction. The author concluded that, to a large extent, the "unknown compounds of *bee origin*" are responsible for the bactericidal properties of honey. This conclusion was made on the basis that pseudo-honey did not differ from real honey in terms of aseptic activity.

We repeated Bogdanov's experiment in a slightly modified form (Isidorov & Bakier, 2011). We fed the three newly created layers of bees from the same family with accepted queens and a very small number of working bees with 50% solutions of sucrose, fructose, and glucose. We conducted the experiment in the summer of 2010 in a no-harvest period, therefore, the bees mainly used syrups. They preferred fructose syrup, then sucrose, with glucose syrup being the least popular. The pseudo-honey taken from capped combs was subjected to GC-MS analysis for the content of acids and sugars. We found with great satisfaction that all three honeys contained royal jelly hydroxy acids. The greatest amounts of which were 10-HDA and 2-DecDA. Based on this data, we queried whether it was possible that these royal jelly acids were the mysterious *bee origin* compounds of honey that Bogdanov wrote about. Our data on the content of many hydroxy acids of royal jelly in various honeys from different countries (Isidorov et al., 2011, 2015), as well as data on the antimicrobial action of these acids (Isidorov et al., 2018), indicate that with the presence of these substances of animal origin, at least part of their inherent antimicrobial activity is associated. Relatively recently, it was found that honey contains other antimicrobial substances of animal (bee) origin of a protein nature, such as defensin-1 (Valachová et al., 2016).

In connection with this, it is worth recalling an experiment in which it was shown that the aseptic properties of honey were enhanced by adding small amounts of royal jelly (Boukraâ et al., 2008). Perhaps in this case we are dealing with the phenomenon of synergy, the meaning of which was grasped by Aristotle ("Metaphysics") in one sentence: *The whole is more than the sum of its components*. It is also worth remembering the antimicrobial peptides, AMP, contained in royal jelly (see Chapter 3.3). In a recent publication (Kwakman et al., 2011), it was reported that one AMP is active against an antibiotic resistant strain of *Bacillus subtilis*.

Analysis of the sugar fraction of these pseudo-honeys also brought interesting results. They indicated that the interconversion (isomerization) of monosaccharides, fructose into glucose and *vice versa*, takes place in the honey sac of bees. For example, as shown in Figure 6.6, when feeding bees with pure fructose syrup, the fructose content of pseudo-honey was approximately 49% and glucose was 43%. Conversely, pure glucose food yielded 38% fructose. In addition to monosaccharides, 17 disaccharides appeared, the content of which was 5–6%. Of sucrose, 0.04% was formed from fructose and 0.1% from glucose. When fed with pure sucrose, it was broken down to 96.5%.

Notably, glucose and fructose, as well as some disaccharides, were represented by several anomers (α- and β-forms). It can be assumed that isomerization occurs under the influence of enzymes from the group of isomerases. It is possible that with their help, the balance of glucose and fructose content, differing in some properties and performing different physiological functions, takes place. For example, glucose crystallizes quite quickly and, therefore, the partial conversion of its excess (noted in the nectar of rapeseed flowers) to fructose is beneficial to bees as it inhibits the crystallization of winter honey stores. In contrast, glucose is essential for the synthesis of trehalose, which is made up of two glucose fragments.

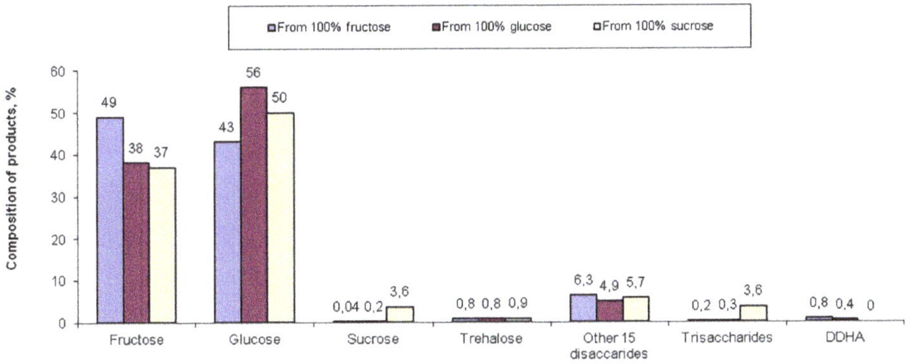

Figure 6.6. Chemical composition of the products obtained when feeding bees with 50% aqueous sugar solutions: fructose, glucose, or sucrose. DDHA - dihydroxyacetone dimer.

Advanced sugar transformations in bees have been previously reported (Barker & Lehner, 1974). According to the data of these authors, the hemolymph of bees fed with sucrose solution contained trehalose, glucose, and fructose. However, when fed with a mixture of sucrose and fructose, twice as much trehalose as glucose was found in the hemolymph and only traces of fructose. The hemolymph of the pure fructose fed bees contained equal amounts of trehalose and glucose and only traces of fructose.

From Figure 6.6, it can be seen that the pseudo-honey obtained from fructose and glucose contained small amounts of a dihydroxyacetone dimer (DDHA), a substance with the formula $C_6H_{12}O_6$, such as fructose and glucose. Therefore, it could potentially be an intermediate product (Fig. 6.7) in the conversion of fructose to glucose and *vice versa*.

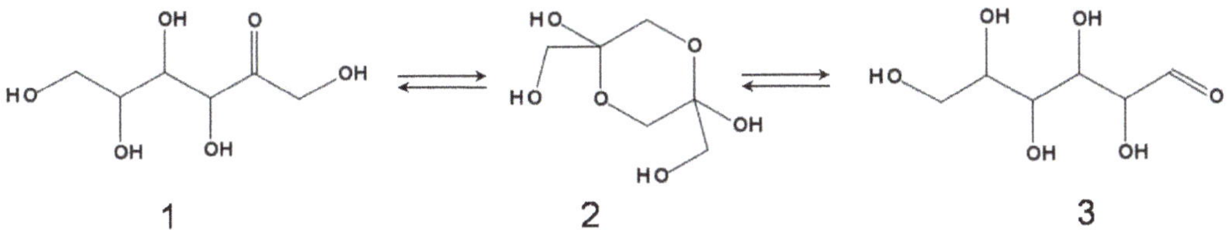

Figure 6.7. Diagram of hypothetical fructose (1) and glucose (3) interconversion with formation of dihydroxyacetone dimer (2) on the transition stage.

The fact that DDHA was not found in pure sucrose honey may not be a coincidence; in an organism with equal amounts of fructose and glucose made from sucrose, there is no need to waste precious energy converting them.

In summary, honey is a complex creation of nature, consisting of many structurally and functionally diverse molecules originating from nectar (or honeydew), pollen, and the bees themselves. If with nectar comes mainly sugars and volatile compounds that determine the aroma of honey, then pollen serves as a source of various types of lipids, phytosterols, carotenoids, and specific proteins, as well as vitamins and minerals. The bees themselves add the secretions of their glands to the nectar they process. In addition to such enzymes as invertases, amylases, glucosidases, and various oxidases, bee glands serve as a source of royal jelly with all the richness in components that it contains, including a set of unique acids and proteins. For nectar to turn into honey, complex and hitherto poorly studied processes must take place. There is still a lot unexplained and incomprehensible about the apparently clear and understandable issue concerning the production of honey. For example, it is not known which enzymes are involved in the transformation of secondary nectar metabolites and what the content of royal jelly ingredients in honey depends on.

6.4. Unifloral honeys and their plant precursors

In most of the scientific publications from recent years on the chemical composition of honey, the authors set a specific goal for themselves, the detection of indicative substances that can be used to explain the belonging of honey to particular varieties. Despite the scientific thread, this task may be of great practical importance. Over the last decade, there have been significant changes to the demands made by consumers for this valuable product. Names such as "forest honey" or "May honey" are no longer attractive. So far, such brands of honey as Bashkir, Omsk, and Altai (honey) are highly respected by Russian consumers. Nowadays, however, there is first and foremost interest in the healing properties that are ascribed to particular unifloral honeys in a multitude of popular publications. As has always been the case, the market carefully analyses and reacts immediately to this demand; therefore, many varieties have emerged, the names of which are associated with well-known medicinal plants or rare and even exotic honey plants.

Unfortunately, it often turns out that such honeys do not correspond to the botanical origin given by the sellers (Lyapunov et al., 2011). Where the market is taken over by vendors, expensive and glorified peach (recommended for arrhythmias and anaemia), coriander (helps dissolve cholesterol in blood vessels), or hawthorn (good for heart diseases) honey may turn out to be ordinary sunflower, rapeseed, or in the best case, linden honey. It all belongs to the category of fake. An even more brazen falsification is the addition of commercial glucose or inverted sugars to honey

The counterfeiting of honey is as old as the world. In the Old Testament, for example, false honey is mentioned from dates and figs. Herodotus also wrote about faked honey. There are also examples of fraud from the relatively recent past.. During the campaign to pass US federal food and drug law (adopted in 1906), one of its promoters, Dr. H.W. Wiley, among other fakes, showed congressmen a jar he had purchased with honey, on which there was a dead bee. The clever scammer thought that seeing a real bee would inspire confidence in the merchandise.

Therefore, the practical side of research on the chemistry of honey concerns not only the detection of biologically active substances in it, but also the protection of the consumer from counterfeit products. An experienced buyer can distinguish real honey and its varieties by smell. However, not everyone is able to recognize the nuances of the aroma of individual unifloral honeys. Objective information about the variety of honey can be obtained on the basis of pollen analysis, a quantitative determination of the pollen content of individual plants in a sample (Persano Oddo & Piro, 2004). Such analysis is performed by specially educated specialists equipped with appropriate equipment and a pollen atlas.

Pollen analysis testifies to the very complicated composition of honey, over 350 ingredients have been detected in its various varieties and this list continues to grow. Unifloral honeys are very different from each other (which is not surprising). Figure 6.8 shows the average composition of the main groups of substances in the 64 samples of some Polish honeys we analyzed. As you can see, aromatic acids are the most numerous in all of them, as well as benzoic acids and its derivatives. The following acids were also identified: salicylic, 4-hydroxybenzoic, vanillic, isovanillic, homovanillic, phenylacetic, 4-hydroxyphenylacetic, 4-hydroxyphenylpropionic, β-phenyl lactic, syringinic, protocatechuic, gallic, phenylpyruvic, veratric, and gentisic acid. In smaller, but not less noticeable amounts, cinnamic acid derivatives are present in almost all honeys, including p-coumaric, ferulic, and caffeic acids.

Each of these acids contributes to the antioxidant and bactericidal properties of honey.

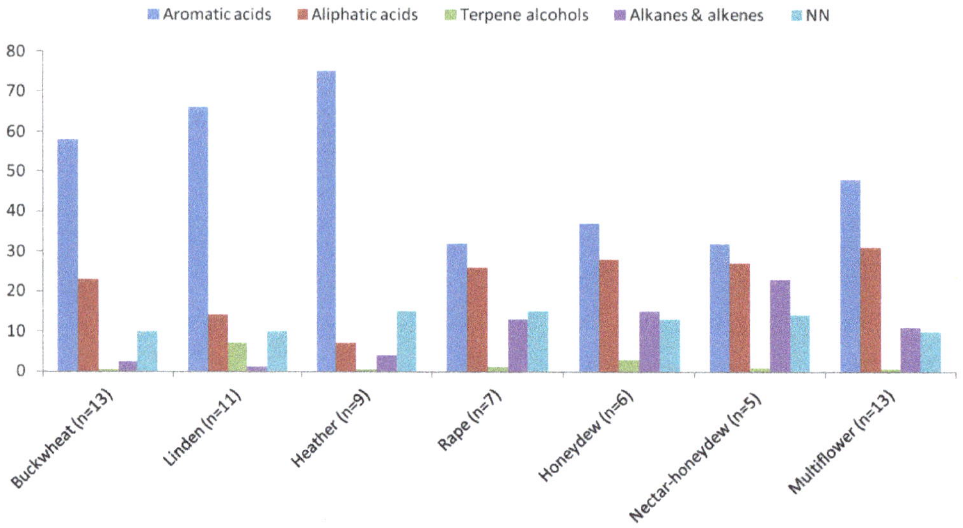

Figure 6.8. Group composition (%) of some varieties of Polish honey, n - number of samples, NN - unidentified substances.

I have not shown flavonoids in this chart, although they are found in honey and many authors write about it. However, it seems to me that they are random admixtures and find their way into the honey from plant precursors other than nectar. In Polish honeys, pinobanksin, chrysin, kaempferol, and quercitin are most frequently detected of all flavonoids. The thought of their random nature suggests that in samples of honeys of the same varieties (and even from the same apiary), they are detected from case to case. For example, out of nine samples of heather honey from southern Poland, we found pinobanksin only in two. Of the 13 samples of buckwheat honey, this flavonoid was present in five, and kaempferol and quercitin, in only two. They are accompanied by 2-phenylethyl caffeate (CAPE). These substances are typical components of poplar-type propolis (Chapter 5.5), which suggests that the flavonoids in honey are external but useful admixtures. Assumptions about the propolis origin of the flavonoids in honey have also been made by other authors (Tomas-Barberan et al., 1993; Frreres et al., 1992; Weston et al., 2000). Tomas-Barberan et al. (1993) described the analysis of flavonoids in honey and found that honeys from the northern hemisphere tended to show higher

degrees of propolis based flavonoids, while equatorial and Australian based honeys were largely devoid of propolis based flavonoids. South American and New Zealand honeys contained flavonoids associated with propolis.

Aliphatic acids are very abundant in honey. All varieties contain lactic acid and its homologues, as well as succinic acid. Some honeys have a high content of higher C_{16} and C_{18} fatty acids, including oleic acid. I will pay attention to the group of ingredients in Fig. 6.8 marked with the letters NN (abbreviation from Latin *Nomen nescio*, which means 'I don't know the name'). These are substances that we have not been able to identify. Their content in this fraction of honey extract was 10–15%.

Many of these chemical indicators have been detected but the situation looks paradoxical; following the report of some authors about finding specific substances for the unifloral honeys they tested, other authors soon discover the same substances in other honeys. An example may be abscisic acid and its transformation product, vomifoliol, as well as dehydrovomifoliol.

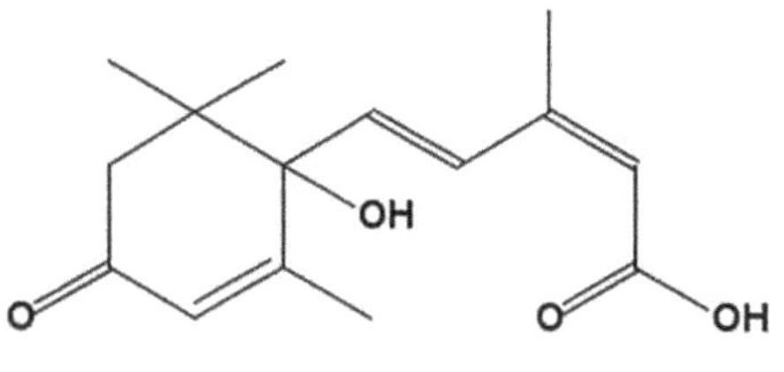

1

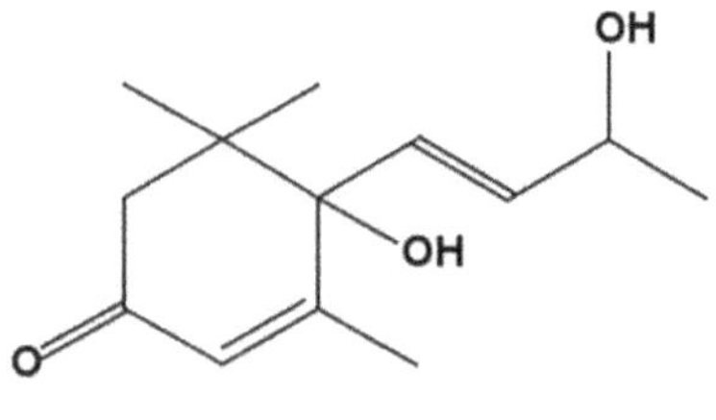

2

Figure 6.9. Formula of abscisic acid (1) and vomifoliol (2).

All three substances were first detected in heather honeys (Guyot et al., 1999; Guyot-Declerck et al., 2000) and it seemed that they could clearly indicate the heather family (*Ericaceae*) as a plant precursor. However, it was later found that they were present in many other unifloral honeys from the nectar of manuka, linden, mint, eucalyptus, or chestnut (Yao et al., 2003; Naef et al., 2004; Jerković et al., 2010; Alissandrakis et al., 2011), as well as in honeydew honey (Jerković & Marijanović, 2010). Abscisic acid is found quite often in a variety of honeys, including acacia, linden, and honeydew. The same is true of other indicators. In addition, their content in honeys of the same variety may vary in different samples. For example, in 13 buckwheat unifloral honey samples, the concentration of homovanillic acid was in the range of 3–25%, that is, it fluctuated 8 times. This makes it difficult or impossible to clearly define a honey variety on the basis of individual indicators. Nevertheless, the task is feasible if we rely on the presence of not one but several characteristic substances in the sample, and at the same time we consider the absence of substances characterizing its type as an analytical parameter. For example, honeydew honey from coniferous and deciduous trees can be distinguished from each other by the absence of 4-hydroxyphenylacetic acid in the latter. This possibility is evidenced by the results of the analysis of the composition of some Polish honeys presented in Table 6.1.

Table 6.1. Potentially “indicator” substances in different honey

Substance	Intervals of the relative content (%) of the substance in honey:							
	acacia	buckwheat	linden	heather	rape	honey-dew*	nectar-honeydew **	multi-flower
Benzoic acid	0.2–1.4	0.1–1.4	0.3–1.4	0.3–1.8	5.2–8.3	0.2–2.5	1.5–5.5	1.1–6.0
Salicylic acid	0–0.3	0.2–0.8	0.2–0.8	0–0.2	0–0.2	1.0–4.6	0.2–1.4	0.1–2.1
4-Hydroxyybenzoic acid	1–3.5	11–40	1–2	trace	1–2	1–6	1.5–5.5	0.1–1
4-Hydroxyphenylacetic acid	-	0.5–2.0	trace	trace	-	7–20	-	0.2–1
β-Phenyllactic acid	0.5–2.5	1.5–13	2–12	57–73	0–1.0	3–10	0.1–3.0	1–32
Homovanillic acid	1.5–4.5	3–25	33–66	1–12	trace	1–8	0.5–3.0	0–1.5
Succinic acid	trace	1.2–8.2	0.3–3.3	0.1–0.8	-	5–18	1–7	1.5–13
Oleic acid	4.5–6.5	0–2.5	0.5–3.5	trace	3.5–5.8	1.5–6.3	3.5–16	1.5–30
Abscisic acid	2.5–4.0	trace	0.5–0.8	0.1–1.5	-	0.1–0.5	0–0.2	trace
4-Hydroxyphenyloethanol	0.2–1.4	0–0.5	0–1.4	0–0.3	2.6-4.5	0–0.3	0.7–1.0	0.05–0.7
Womifoliol & dehydrowomifoliol	-	0.5–2.6	0–0.5	4–11	-	-	trace	0–0.3
Metyl syringate	0.9–1.5	trace	0–1.0	-	10–16	trace	0.9–4.8	0.3–8

* *From coniferous honeydew.* ** *From deciduous honeydew.*

As we can see, individual varieties of honey are characterized by their own profile (set of substances), which make it possible to detect its botanical origin. Most often, this profile is obtained by analyzing (e.g. by gas chromatography) extracts of honey with various solvents or with the use of appropriate adsorbents. However, such a procedure is very laborious and time-consuming. The analysis is much faster using solid phase microextraction (SPME), which was discussed in chapter 5.4. This method combines two stages, sampling and preparation for analysis, which greatly simplifies and shortens the analysis time. For this reason, in recent years, SPME has been increasingly used in the analysis of honeys (Soria et al., 2009; Aliferis et al., 2010; Bianchi et al., 2011), including Polish ones (Wolski et al., 2006; Plutowska et al., 2011).

It is too early to talk about the use of data obtained by this method for the detection of honey varieties because obtained experimental material is not sufficient[20]. Notably, SPME and classical extraction give different profiles. When using the first method, only volatile and low-polar substances that determine the honey aroma are detected. Most of these compounds are lost during classical extraction with solvents or adsorbents. Conversely, the classical method makes it possible to determine the polar substances on which mainly the antioxidant and antimicrobial properties of honey depend. This means that the two methods are complementary and should be used when detecting the overall chemical profile of honey.

In this chapter, we mainly discussed the chemical composition of Polish honeys, although we also examined many varieties of honey from other countries, including Russia, Croatia, Austria, France, Scotia, Spain, USA, Morocco, and Egypt. All samples were true honey and we never encountered any falsification. We also tested the Manuka nectar honey, MGO™ 400+, which comes from New Zealand and is highly praised. Manuka honey is characterized by a strong bactericidal effect due to the presence of methylglyoxal (MGO). This compound is gradually formed during the maturation of honey from the dihydroacetone present in manuka nectar. Therefore, the content of MGO in honey can be different, which is reflected in the trade name of the product. For example, the name MGO 400+ means that the content of methylglyoxal is not less than 400 mg/kg (and it can even reach 1000 mg/

[20] *I can only point out that the composition of the volatile secretions of various honeys is very different. As a result of the SPME analysis of six unifloral honey samples from Russia and six samples from Poland, we identified over 120 substances. However, only 13 of them were common, i.e. present in the secretions of all 12 honeys.*

kg). Honestly, I was not particularly delighted with the chemical composition of this honey. In one of the scientific centres of Białystok, its bactericidal properties were also tested. The conclusion was that it does have such properties but it does not stand out very well.

It is possible that we were not lucky and in the jars purchased through the distribution network with the inscriptions: 'Health Manuka Honey MGO 550+/100% Pure NZ' there was an ordinary fake. As a result of the high cost of manuka honey, an increasing number of products, which are now being labelled as such around the world, are counterfeited or falsified. It has recently become known that with an annual production of this honey in New Zealand of about 1,700 tons, its supply on the market is about 10,000 tons, including 1,800 tons in England. In governmental agency tests in the UK between 2011 and 2013, a majority of manuka-labelled honeys sampled lacked the non-peroxide anti-microbial activity typical of manuka honey.

In a word, this honey is distinguished mainly by its price. One kg of manuka honey costs as much as about 16 kg of the most expensive Polish heather honey. Moreover, manuka honeys with a high content of methylglyoxal (MGO™-250+ and MGO™-550+) cannot be eaten, since this compound demonstrated a mutagenic effect when tested on animal cell cultures and its mutagenicity is significantly enhanced in the presence of H_2O_2 (IARC Monographs, 1991).

However, there are no contraindications against its external use for medicinal purposes. This application of manuka honey will be discussed in the next section.

6. 5. Medical-grade honey

In traditional medicine, among all peoples familiar with honey, it has always occupied an honourable place. People knew the medicinal properties of many types of honey, such as the diaphoretic action of linden honey used for fever. One of the main medicinal uses of honey since the time of Hippocrates has been the treatment of wounds (Hippocrates. Internet Archive of the Classics), burns, and bedsores. Thus, for thousands of years and until recently, even during the First World War, honey was used in the treatment of wounds. However, in the middle of the twentieth century, antibiotics, sulphonamides, and hormonal drugs replaced honey from the standpoint of a medicine. Honey began to be viewed as a healthy and easily digestible food product that helps to restore the health of the patient and maintain it in a healthy person. Probably, only a nutritionist would prescribe honey but would not send the patient to the pharmacy with a prescription. Many doctors and specialists still consider honey a folk remedy without scientific evidence of its use in modern medicine.

The last decades have been characterized by a renewed interest in honey, mainly due to its antibiotic properties. This is due to a circumstance known and very unpleasant in its consequences, the appearance of strains of many types of pathogens that are resistant to antibiotics. Interestingly, for the emergence of a microbe strain resistant to any new antibiotic, a fairly short time of its widespread use is sufficient. In relation to honey, as well as to propolis, microbes have not been able to acquire this quality for many centuries and millennia.

Although some attribute the antimicrobial properties of honey to the presence of hydrogen peroxide, while others with methylglyoxal, and others with defensins, they are due to a unique and variable set of various chemical factors. It is this circumstance that counteracts the development of resistance in microbes, while these organisms would find a remedy for any single factor. For example, one can easily imagine a strain of a certain fungus or bacteria capable of producing catalase that effectively breaks down H2O2 or other enzymes that oxidize methylglyoxal.

Difficulties in healing wounds and burns are primarily associated with the occurrence of infections and with the slow regeneration of integumentary tissues. Being non-toxic to human tissues and possessing antiseptic and regenerative properties, honey serves as the best way to overcome them. Therefore, the medical use of honey is mainly in this direction. Here, honey has completely restored its place in official medicine. The therapeutic properties of honey in the treatment of wounds and burns, as well as the prevention of postoperative infections are described in detail in a number of review articles (Moore et al., 2001; Boukraâ, 2013; Oryan et al., 2016). Without going into detail, these include:

- antimicrobial action,
- anti-inflammatory action,
- prevention of oedema and pain relief,
- reduction of odour from the wound,
- promotes rapid healing with minimal scarring.

Honey medicated dressings are much less painful than dressings with propolis, which are characterized by strong adhesion but also used successfully in the treatment of wounds and burns.

From a medical point of view, honey is harmless. Allergic reactions rarely occur and, in most cases, they are associated with the presence of specific types of pollen. However, this can be easily remedied by passing the honey through a fine filter. In theory, cases of botulism, a form of paralysis caused by a neurotoxin produced by the microbe *Clostridium botulinum*, are possible. These anaerobic bacteria are widespread in the environment and are also found in honey, although not so often; they were present in 5% of the 2000 honey samples examined. The content of *C. botulinum* spores in honey contaminated by them is also not high and ranges from 18 to 60 spores/kg of honey (Midura et al., 1979). When using honey dressings on wounds, anaerobic conditions are created and the possibility of germination of pathogen spores arises. However, to date, there have been no published reports of cases of such patient toxicity.

Not all types of honey are approved for medical use and a new category of beekeeping product is now emerging, medical-grade honey (MGH). This product has a number of requirements:

- sterility (absence of spores of bacteria capable of forming vegetative forms when honey is diluted with body fluids during a treatment procedure);
- speed and wide spectrum of action on microorganisms, including antibiotic-resistant strains of bacteria;
- the presence of regenerative properties;
- batch-to-batch reproducibility of the composition of characteristics that determine the therapeutic effect;
- low cost.

Interest in this kind of products is evidenced in the period from 2002 to 2015. In that period, 14 patents were published for their production. However, at present, as far as I know, only a few are used in medical practice and the most famous are Manuka and Revamil® honeys. None of these are inherently sterile and, in general, one can doubt that such honeys exist. We have already said that bees bring microbial spores together with nectar and pollen and beekeepers do their part. In recent studies of Turkish honey of various origins, only seven of 54 tested samples failed to detect microbes (Borum & Gunes, 2018). In mature honey, due to the low water content, only spores can exist. However, when it is diluted with body fluids during a treatment procedure (applying dressings soaked in honey), conditions are created for the formation of vegetative forms of microbes capable of reproduction. Therefore, honeys used in the manufacture of licensed products for the treatment of wounds are sterilized by gamma radiation. Fortunately, it turned out that at a dose of 10, 20, and 30 kGy, the antibiotic properties of honey are not lost, although, there is a slight decrease in the content of antibacterial low molecular weight peptides, in particular defensin-1 (Horniacková et al., 2017).

The mechanism of action of Manuka honey and a product called Revamil® are different; if the therapeutic effect of the first of them is determined by the high content of MGO (Kwakman et al., 2010; Watson et al., 2017), the antibiotic effect of the second, almost devoid of this component, is associated with hydrogen peroxide and defensin-1 (Kwakman et al., 2011). According to Kwakman et al. (2011), honey from manuka nectar contains practically no defensins and when it is diluted, hydrogen peroxide is not formed, although the latter seems surprising to me (this means that Manuka honey is completely devoid of the enzyme glucosidase, which is hard to believe.). Revamil® is made from honey, about which little is known. It is produced in the central part of the Netherlands by Bfactory Health Products (Rhenen). It is reportedly obtained under tightly controlled conditions from bee colonies housed in greenhouses but the plant precursors are kept secret.

The action of these honeys is also different. Revamil® has a fast action (within two hours) against antibiotic-resistant strains of *Pseudomonas aeruginosa*, *Staphylococcus epidermis*, *Enterocossus faecium*, and *Burkholderia cepacia* but does not quickly act on methicillin-resistant *S. aureus* (MRSA) and *E. coli*. The rate of action on all tested bacterial species, including MRSA, is reduced to two hours after the initial honey is enriched with 75 µM of the synthetic bactericidal peptide BP2. Manuka does not act quickly on *B. subtilis* but in 24 hours it kills all tested bacteria, including MRSA (Kwakman et al., 2010).

Other products are also included in the category of medical-grade honey, except for the two named and recognized ones. These include SurgihoneyRO™ from England and Israel's LifeMel. Their origin is no less mysterious than the honey for Revamil®. The first is said to be a bioengineered product that has undergone a patented process to enhance its natural antimicrobial activity, derived from the sustained release of hydrogen peroxide. Regarding the LifeMel, I have strong suspicions that this is a mixture of honey with extracts from 16 medicinal plants, including eleuthero, lemon balm, calendula, and dandelion, or a beekeeping product, which will be discussed in the next chapter.

CHAPTER 7
Not from nectar or honeydew, but still honey

7.1. Required explanation

To avoid confusion and misunderstandings, at the very beginning of this chapter it is necessary to explain what kind of beekeeping product it will talk about. It was superfluous in the Polish edition of the book, because in Poland, as in some other Eastern European countries, it has been known for a long time. In the scientific literature, the name *herbhoney* (HH) is used for it, and here a misunderstanding may arise that you want to avoid. On the Internet, you can find different recipes for the preparation of so-called *herb-infused honey*, or just *herb honey*, for example, this:

- fill a clean quart jar a little less than halfway with dried herbs and spices,
- pour in your honey and watch as it slowly finds its way to the bottom,
- put a lid on the jar and place on a sunny windowsill,
- turn the jar over at least once per day.

As you can see from the above recipe, this product is prepared at home, while we will get acquainted with the product made in the hive, that is, by its inhabitants, bees. The product discussed in this chapter is prepared in a completely different way. As an example, I will give a method of preparing one of many HH varieties.

In late spring, young soft shoots are harvested at the ends of pine branches, on which needles have not yet developed. These shoots are covered with sugar in a weight ratio of about 1:1 and left under a lid at ambient temperature (at which it begins to secrete juice quickly) until required. Usually this is in July-August, that is, after the main honey harvest, although unfavourable conditions (frequent rains, lack of nectar or honeydew) can force the beekeeper to start doing this earlier. The resulting dark green liquid is separated from the pine branches, and sugar dissolved in boiling water is added so that a 50% sugar solution is obtained. This syrup is served in feeders under the covers of the hives and the bees willingly take it. The bees are given the time required to process the syrup (similar to processing flower nectar or honeydew), after which the finished product is extracted. As you can see, it does not form on the kitchen table, but in the hive, with the direct participation of bees.

Instead of sugar, some beekeepers use reserves of unused honey, diluting it with water to the required concentration. Not only pine shoots, but also other plant material is used for these purposes. In Poland, large-scale producers are the big beekeeping enterprises such as Apipol-Kraków and Sądecki Bartnik®, which offer six to ten different kinds of HH depending on the seasonal availability of plant raw material including the green parts of common nettle and thyme, and hawthorn flowers, as well as aloe and chokeberry juice. All these HHs are in great demand, despite the higher cost compared to nectar honeys.

7.2. Honey or not honey? That's the question

Before starting to write this section, I decided to look again at what is written on the Polish beekeeping forums about HH, and how the beekeepers themselves relate to it. The last time I did this was 5–6 years ago, and I wondered what changes had occurred. As it turned out, Polish beekeepers have not stopped exchanging opinions about HH, but the tone of the discussions has changed markedly. Previously, almost exclusively swear words were used when referring to HH ("Consumer deception!", "Falsification!", "Shame!", etc.), accompanied by appeals and demands to "Ban!" Anger was mainly caused by the use of the word "honey" in the name of this product. Recently (2015–2020), quite sharp statements on this topic have also appeared, but in general the discussion is being conducted in a calmer tone.

What questions do those beekeepers who are interested in HH ask of their colleagues? Here are some of them:

- Why is it better than honey mixed with a decoction of herbs?
- Does this "honey" contain any of these herbs?
- How long can HH be stored? Would it not happen that, without being used or sold, it quickly becomes fermented?
- How do bee colonies react to HH? Do they harm the bees?

I will try to give answers to these questions, based on the results of my own research, as well as on the reports of Polish scientists. But at the beginning I will say a few words about the history of HH and touch on the issue of terminology.

In all likelihood, the idea of producing HH belongs to N.P. Ioirish Ph.D. He began his experiments in 1939 at a taiga apiary in the Primorsky Territory off the Russian Pacific coast. The goal of this scientist doctor was not to increase beekeeping production or to increase the economic indicators of the industry, but to obtain new medicinal products that would combine the beneficial properties of honey and medicinal herbs. What occupied his mind was whether it would be possible to introduce biologically active components into honey and thus increase its medicinal value.

Figure 7.1. The creator of the herbhoney military doctor N.P. Ioirish

Indeed, the value of honey should increase if it contains more vitamins, for example. These compounds, important for humans, are also present in nectar, but their content is so low that they do not have any noticeable physiological effect. On the other hand, the vitamin content in, for example, rose hips and red carrot juice is very high. After feeding the syrup of carrot juice (Ioirish named it an “artificial nectar”) and sugar to the bees, “carrot honey” appeared for the first time, the analysis of which showed a high content of vitamins: B_1, B_2, C, P and PP, as well as pantothenic acid. The daily requirement of an adult for vitamins is covered by what is contained in 25 g of carrot honey. In addition to that, carrot honey contains more phosphorus, calcium and iron salts than a person needs. Ioirish also drew attention to other plant “raw materials” rich in vitamins and available in unlimited quantities – pine, fir and spruce needles (by the way, he found that it is better to use winter needles, because in spring and summer their vitamin content decreases). Ioirish, as a doctor, was also interested in the possibility of introducing medications into honey, such as antibacterial sulphanilamides (streptocid), terribly bitter quinine and calcium chloride. His experiments were successful, but I will not hold the Reader’s precious attention on them.

Now a little about terminology. At the end of the last century and at the beginning of the current one, a heated discussion about the name flared up among Polish beekeepers or, more precisely, a wave of protests arose against the use of the phrase "ziolomiod" (this is what Ioirish's HH is called in Poland). Ziolomiod is a compound word, the first part of which comes from the word "ziele" – a medicinal herb – and the second part "miod" is translated as "honey". I cannot say that these protests are absolutely unfounded, since the concept of "honey" is well defined. For example, the order of the Minister of Agriculture of Poland defines honey as follows:

- "a natural sweet product produced by *Apis mellifera* bees by combining plant nectar with their own specific substances, or secretions of living parts of plants, or secretions of insects that suck the juices of living parts of plants, set aside, dehydrated and left to ripen in combs."

For my taste, this definition is awkward, but in essence does not differ from those adopted in other countries.

But, on the other hand, I cannot agree with questions and statements of this kind: "Why should this product be called herbhoney? After all, it has nothing to do with honey!" The feature common to honey and HH is that they are produced by the bees themselves, and I see no difference in the way the "raw materials" are processed. Yes, their precursors are different, but both are given to us by wildlife, and not by chemical industry enterprises, for which oil, coal and other minerals serve as raw materials. Moreover, both plant nectar and sugar syrup containing an extract from medicinal plants are subject to long-term processing by bees. However, judging by the discussion on the forums, the majority of Polish beekeepers now have nothing against this name, and they insist on only one thing: the consumer must be informed about the origin of the product.

Indeed, why is nobody confused by the phrase "artificial honey"? Well, the latter has absolutely nothing to do with honey and bees! You can probably think of some other name for HH, but here's the catch: what to do with a product prepared by bees not from sugar syrup and plant extracts, but from real honey and the same extracts? Is it possible to graciously grant it the right to be called "honey", or should it also invent something? Agree, Reader: absolutely precise and unambiguous definitions do not exist, and we easily put up with this. Sometimes we even prefer to use the wrong, but familiar, and nothing happens, we live! Remember at least two names for the honey bee: the generally accepted *A. mellifera* and the more accurate, but not established, *A. mellifica*. In the end, the most important thing is not the name, but the product itself. And here a legitimate question arises: what new valuable qualities does this product have? What does it give to the allegedly "deceived" consumer?

7.3. Herbhoneys are different from honeys but not for the worse

I will start with the answers to the questions that the beekeepers themselves put on the forums and, may their participants forgive me, I will use the answers and remarks of some of them without naming them.

First, the question of *why herbhoney is better than honey mixed with herbal decoction* can be answered simply: it tastes better. And taste is one of the main characteristics of any food product. A decoction of nettle is unpleasant in taste and smell, and can spoil your favourite honey. But HH, made by bees from sugar syrup with the addition of nettle juice, smells good and delicious. But this is not the most important thing either. It is more important to answer the second question: *Does the herbhoney contain any of the herbs used to make syrups?*

Of course it does, and it could not be otherwise: no matter how rich in chemical composition the nectars of flowers, extracts from the same flowers, leaves and fruits contain much more biologically active substances. For example, there are always flavonoids in the green parts of plants and these are transferred to HHs (Fig. 7.2). In the case of pine HH, this is a catechin that has a vaso-strengthening effect. In thyme HH, there are chrysin, galangin, naringenin, luteolin and other flavonoids; in nettle and hawthorn HH, there are kaempferol and quercetin. On the other hand, no flavonoids were found in Polish monofloral honeys from Sądecki Bartnik®. Thyme HH contains thymol, which is a very powerful antimicrobial substance that surpasses any of the flavonoids in action.

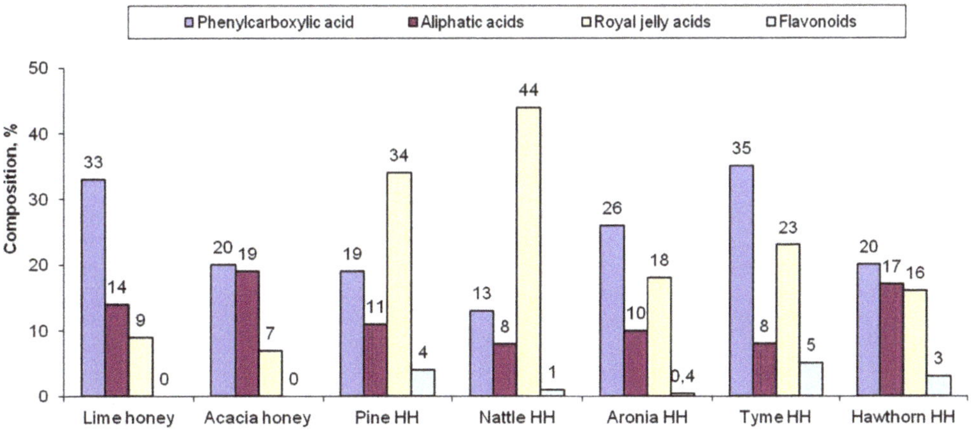

Figure 7.2. Group composition of some compounds in lime and acacia honey, as well as in five herbhoneys.

In the available scientific literature, I did not find any reports devoted to a detailed comparison of the chemical composition of "artificial nectars" and HHs obtained from them, so I instructed one of my students to conduct this kind of research. As a result, interesting differences were found in the composition of the initial and final product (Zalewski, 2017). These features are visible from Table 7.1.

Table 7.1. Relative composition (% of TIC) of "artificial nectars" (AN) and herbhoneys (HH) obtained from them.

Components	Goatweed HH		Pine HH	
	AN	HH	AN	HH
Rojal jelly acids	-	28.38	-	17.7
Amino acids	3.64	-	-	-
Phosphoric acid and phosphates	0.15	-	-	-
Di- and trisaccharide	52.4	-	51.8	-
3-Caffeoylquinic acids, including:	12.86	-	-	-
– (Z)-neochlorogenic acid	0.32	-	-	-
– (E)-chlorogenic acid	12.54	-	-	-
Hydroxycinnamic acids, including:	-	12.37	4.6	10.0
– p-coumaric acid	-	5.81	2.4	6.5
– (Z)-caffeic acid	-	0.12	-	-
- (E)-caffeic acid	-	4.55	0.6	1.5
– (E)-ferulic acid	-	1.80	1.6	2.0
Quinic acid	11.06	-	-	-
Resin acids	-	-	28.3	-
Sterols	-	0.39	-	1.4
Flavonoids, including	-	0.48	25.0	14.2
– epi-catechin			1.5	-
– catechin			2.4	-

We were no longer surprised by the appearance in HHs of hydroxy acids characteristic of royal jelly (this was discussed in the previous chapter).[21] The disappearance of di- and trisaccharides, which are decomposed by glycosidase contained in the saliva of bees, is also quite understandable. It can be assumed that a

[21] *There is no doubt that HHs also contain other constituent components of royal jelly, such as antimicrobial peptides, AMP. I cannot unequivocally answer the question why there are more of these compounds in HHs than in monofloral ones. Could it be that more young nursing bees with well-developed glands that produce royal jelly are included in the processing of the "home-delivered" syrup?*

number of hydroxycinnamic acids, sterols and flavonoids also owe their appearance in HHs to glycosidases, since in plant tissues these compounds are present mainly in the form of glycosides. But the disappearance of some other compounds during the processing of the original "artificial nectars" by bees attracts special attention.

In the "artificial nectar" from goutweed (*Aegopodium podagraria*), 12 different amino acids were found, but none of them passed into HH. The same is observed in the case of phosphorus compounds. And then an analogy comes to mind with aphids, which effectively extract amino acids and peptides diluted in plant juices and excrete excess sugars. On the other hand, when bees process "artificial nectar" from goutweed, the chlorogenic acids characteristic of it disappear. By chemical nature, they are esters of quinic acid with (*Z*)- and (*E*)-isomers of caffeic acid. These isomeric acids are absent in artificial goutweed syrup, but appear in the herbhoney obtained from it! This suggests that during the processing of syrup by bees, hydrolysis of chlorogenic acids occurs (Fig. 7.3).

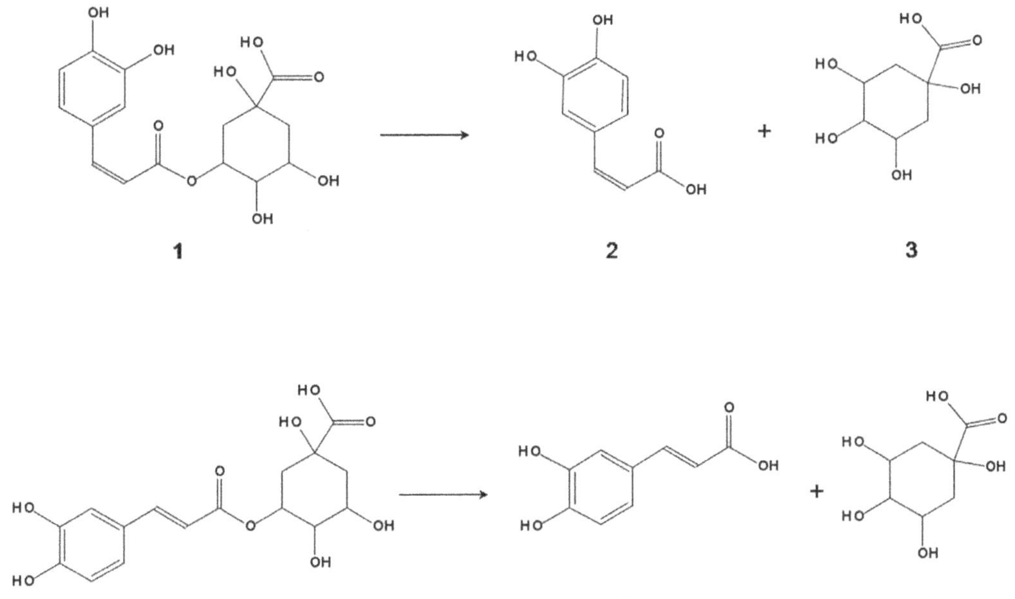

1 2 3

4 5 3

Figure 7.3. Hydrolysis of neo-chlorogenic acid (1) with the formation of (Z)-caffeic acid (2), as well as chlorogenic acid (4) with formation of (E)-cafeic acid (5). In both cases, quinic acid (3) is also formed.

This, in turn, means that not only glycosidases but also carboxylesterases enter the honey sac of bees. It is these enzymes that catalyse the hydrolysis of esters. The presence of esterases in the midgut of bees has been known for a long time; however, I have not heard of their participation in the processing of nectar. But here is the question: why in goutweed HH is there only part of the hydrolysis products of chlorogenic acids, namely, (*E*)- and (*Z*)-caffeic acids, but no quinic acid? There was quite a lot of it already in the original syrup itself (11.06%), and the hydrolysis of chlorogenic acids should have approximately doubled its content. Its complete disappearance remains a mystery, as does the complete disappearance of resin acids from pine "artificial nectar" (Table 7.1).

Royal jelly hydroxy acids, hydroxycinnamic acids and flavonoids, as we know, have antimicrobial effects. Therefore, we were not surprised by the results presented in Table 7.2, according to which HHs are distinguished by higher antimicrobial activity in comparison with "true" or "genuine" honeys (I will explain: the lower the MIC value, the stronger the antimicrobial effect of the tested honey against the specified type of bacteria).

Table 7.2. Antimicrobial activity of herbhoney and genuine honey extracts (Isidorov et al., 2015).

Product	Minimal inhibitory concentration (µg/ml)*			
	Bacillus cerreus	*Staphylococcus aureus*	*Staphylococcus schleiferi*	*Candida albicans*
Herbhoney (HH)				
Pine HH-1	0.625	1.250	-	1.250
Pine HH-2	0.625	1.250	0.625	1.250
Hawthorn HH-1	0.078	0.625	1.250	0.078
Hawthorn HH-2	0.078	1.250	2.500	0.625
Nettle HH	0.078	0.625	5.000	0.078
Thyme HH	0.078	0.312	1.250	0.039
Aloe HH	0.039	0.625	1.250	0.020
Black chokeberry HH	0.039	1.250	1.250	0.625
Genuine honey (GH)				
Heather (GH)	0.215	1.250	0.156	1.250
Buckwheat (GH)	0.625	1.250	0.313	0.937
Multifloral (GH)	-	2.500	-	1.250

**Escherichia coli: MIC>5.0 µg/ml for all kind of HH and GH extracts.*

HHs are also distinguished by a very high content of micronutrients. These are minerals that play a huge role in metabolic processes in the body. For the most part, they are enzymes, catalysts for biochemical reactions. A lack of micronutrients leads to serious diseases. There are also minerals in honey, but in very small quantities. For example, their total content in honey from flower nectar usually does not exceed 5 mg/kg, and in honeydew honeys 15–20 mg/kg, with the main part being macro components, magnesium and calcium. Compared to these two varieties of "true" honey, HH is a treasure trove of micronutrients.

Studies of the elemental composition of six Polish HHs carried out for us at the Department of Laser Physics and Spectroscopy of the University of Grodno showed that the total content of only eight micronutrients (out of 27 found in samples) was 40–116 mg/kg. As seen in Fig. 7.4, thyme HH stands out among others for this indicator: it contains the most micronutrients (the main ones are iron and zinc), and almost the same amount of macro components (potassium and calcium) as in nettle HH.

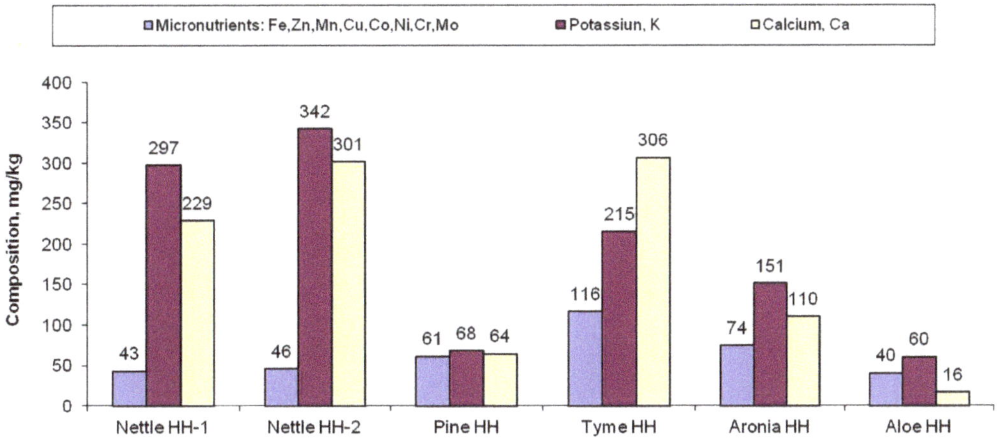

Figure 7.4. Content of mineral components in herbhoney from Apipol-Kraków nettle-1, thyme, chokeberry and aloe products, and Sądecki Bartnik® nettle-2 and pine products.

HHs produced by Apipol-Kraków (10 types) have been tested for their compliance with Polish honey regulations (Juszczak et al., 2009). Here are some of these indicators. The water content was in the range 15.8–18.3%, the diastase number was from 13.1 to 46.2, the sucrose content was 0.4% to 6.4% (the exception was raspberry HH which had 24.8% sucrose) and total acidity was 16.5–30.5 meq/

kg. The content of 5-(hydroxymethyl) furfural, HMF, slightly exceeded the norm only in the case of HH from chokeberry (37.3 mg/kg) and mint (33.0 mg/kg); in the rest it was in the range of 5.6–18.9 mg/kg. HHs provided to us for research by Sądecki Bartnik® were also characterized by similar indicators. The water content in five types of product (pine, hawthorn, nettle, chokeberry and aloe HHs) was in the range of 15.5–17.1%, diastase number was from 6.5 to 13.9, sucrose content was 0.4–1.6%, and HMF content was 6.7–28.5 mg/kg. That is, all these products meet the requirements for "true" types of honey.

The next publication by Polish scientists (Socha et al., 2009) reported a high content of phenolic compounds in samples of HH, including flavonoids, and a high antioxidant capacity of HH. Unfortunately, I came across only one work which presents the results of determining the antioxidant capacity of honey and HH. According to Lithuanian authors (Baltrušaitutė et al., 2007), HHs obtained using extracts of pine, nettle and birch showed higher antioxidant properties than rapeseed, linden and other honeys.

One of the works cited above, Juszczak et al. (2009) provides an answer to another question posed by one of the forum participants: *How long can herbhoney be stored? Would it not happen that, without being used or sold, it quickly becomes fermented?* The answer is: no less long than any monofloral honey, and there is no danger of its fermentation (of course, subject to the usual requirements for honey maturity and storage). This is due to the low water activity in honey.

Water activity (aw) shows how strongly the water in a product is bound to other compounds it contains. The sugars in honey (glucose, fructose and trehalose) bind strongly and retain water. Therefore, with its seemingly high content in honey (18–20%), water is inaccessible to microbes. For pure water, aw = 1.0. In different honeys, aw is 0.5–0.6. Inhibition of the development of one of the most "hardy" bacteria, Staphylococcus aureus, occurs at aw = 0.86. For microscopic fungi that cause honey fermentation, this border lies near aw ≈ 0.70. Loss of antibacterial properties of honey can occur with prolonged contact with humid air. Honey "sucks" moisture out of the air, the aw value rises and honey becomes an excellent medium for the development of microbes.

And finally, one more very important question: *How do bee families react to herbhoney? Does it harm the bees?* We will find comments on this in the works of Ioirish, and the answer, supported by experiments and observations, comes from Polish scientists and participants in beekeeping forums. N.P. Ioirish (1976) wrote: "Some 'artificial nectar' has a very beneficial effect on the bee colony. Some of the substances that make up the 'artificial nectar' stimulate the queen bee to lay eggs even in late autumn, others activate the bees to vigorously rebuild the honeycomb."

Polish scientists have evaluated the possibility of feeding bees with herbal extracts in order to increase their viability (Pohorecka & Skubida, 2002). They fed the bees with syrups with extracts from five plant species: calendula (*Calendula officinalis*), echinacea (*Echinacea angustifolia*), plantain (*Plantago lanceolata*), dandelion (*Taraxacum officinale*) and nettle (*Urtica dioica*). Bee families who were given pure sugar syrup served as controls.

I will briefly present the main results of this interesting experiment. The bees took the extract syrups as well as the pure sugar syrup, with the exception of the calendula syrup. Perhaps for this reason, the greatest loss of body weight and the highest mortality rate in the first 20 days of the experiment was noted in this group. The lowest mortality rate and body weight loss was recorded in bees fed with syrups based on nettle and echinacea, and their mortality rate was lower than in the control group (in the case of nettle extracts, mortality was 12% less). Nettle extract also had a positive effect on development of the bees' mandibular glands. The conclusion of the authors of the experiment is that feeding bees with syrup containing extracts from stinging nettle and echinacea has a beneficial effect on their body. This is the result of laboratory observations, and what about natural conditions? I cannot refrain from citing only three "testimonies" without any changes:

"On Saturday I pumped out the nettle herbhoney – to me, how delicious it is! – not to be compared with syrup with nettle decoction, which has a nasty taste and smell. On occasion, I noticed: the 'girls' rebuilt the honeycomb like crazy, and the queens laid eggs, as if they had lost their measure."

"I don't sell herbhoneys because they are too valuable for my family and take a long time to prepare (collecting, drying flowers, herbs and fruits, making syrups, etc.). However, it is worth getting them for yourself in order to enjoy good health, and they are useful for bees too."

"I once gave the bees a decoction of elderberry flower to process. What I took out of the hive – delicious, I did not sell to anyone, I ate it myself."

Please note that these statements are taken by me from beekeeping forums, where beekeepers, that is, people who know a lot about honey, exchange experience and knowledge with their colleagues. And their observations concerning the influence of HH on the health of bees deserve attention. Besides, who, if not beekeepers, understands the taste of honey!

All of the above can be summarized in the following theses which, however, reflect only the personal opinion of the author:

1. Yes, herbhoney is not *genuine honey*! This is another valuable beekeeping product, which requires additional and not inconsiderable effort. Therefore, it costs almost as much as the most expensive Polish honey, heather.
2. HHs are distinguished by a rich chemical composition and some of them surpass honey in such important indicators as antimicrobial and antioxidant effects.
3. There can be no talk of any deception of the consumer. By selling HH, the beekeeper has the right to claim that he or she is offering the highest quality product, and not some kind of surrogate. It is important that it is prepared in good faith, in compliance with all hygiene rules and on the basis of good-quality materials (including high-quality sugar).
4. The effect of HH on bees should be investigated more thoroughly, in particular, the possibility of using it as medicinal and winter food.

In conclusion, a few words about the medical use of HH. One of the most important requirements for medical-grade honey (see section 6.5) is the reproducibility of its properties, that is, the chemical composition. Such reproducibility can hardly be expected from nectar or honeydew honeys obtained in the traditional way due to natural variations in the composition of their plant precursors. As a result, regular honeys are usually different from batch to batch. However, it is not difficult to obtain an artificial sugar syrup with the desired properties from standardized plant materials and feed it to bees. The medicinal properties of the product obtained for the treatment of wounds and burns can be purposefully improved by adding plant extracts with regenerative properties to the syrup. This syrup can be gamma-irradiated before serving it to the bees to kill the spores of microorganisms. And, of course, it is desirable to keep the bees isolated from external sources of infection (as well as from random sources of plant nectars, which can lead to a deviation of the resulting product from the specified parameters). I have strong suspicions that the mysterious Revamil, whose description states that "it is standardized, medical-grade honey, produced under controlled conditions in greenhouses, by Bfactory Health Products, Rhenen, The Netherlands" is produced just in this way. That is, it is nothing more than HH.

CHAPTER 8
Beebread – Food of the Gods

8.1. Pollen load and beebread, are they one and the same?

I confess that it is completely incomprehensible to me why, judging by the number of publications, so little attention is paid to beebread in the scientific literature. Even in specialized scientific journals such as *Apidologie*, *Journal of Apicultural Research* and *Journal of Apicultural Sciences*, it is possible to find only a few articles in which at least it is mentioned. And this despite the fact that beebread certainly belongs to the oldest biologically active additives (BADs). Truth be told, there have been three review articles published on this product in the last three years (Kieliszek et al., 2018; Mărgăoan et al., 2019; Khalifa et al., 2020).

Since ancient times, people have been faced with a lack of vitamins, some amino acids, mineral components and other substances. The body's need for them is small, but the lack or absence is detrimental to health. This type of nutritional deficiency is not necessarily associated with poverty and starvation; it can be seasonal or endemic. Therefore, people have tried to overcome it on the basis of empirical experience, selecting substances and products that serve not so much to satisfy hunger as to improve the health of the body. Information about such products is contained in numerous sources that have survived to this day on ancient Egyptian, ancient Chinese, ancient Indian and ancient Greek medicine. If we talk about beebread, then in some ancient sources it was assigned such self-explanatory epithets as "Spring of Youth" or "Food of the Gods". The ancient Greek physician Hippocrates also paid attention to the healing effect of beebread. The knowledge of the ancients was not lost in later times, as evidenced by the books of numerous doctors who paid great attention to health promotion. Among the means contributing to this, beebread occupied one of the leading places.

What is the situation now? According to the testimony of James Higgins of the American Apitherapeutic Society, beebread is included in the diet of almost all Olympic-level athletes, as its intake improves tissue absorption of oxygen, helps athletes overcome forced loads, and also quickly restores strength. It is a great natural and safe anabolic, that is, it helps to increase muscle mass and stabilizes the endocrine glands.

This is not to say that beebread is a publicly available product. Rather, it refers to exclusive products. A much more common and readily available dietary supplement is pollen load: pollen collected by bees and delivered to their nest in baskets on the last pair of legs.

To collect pollen load, beekeepers use special devices installed at the entrance, pollen catchers incorporating a grid with 4.5 to 5 mm holes. The bee arriving with the prey is forced to pass through these holes, while part of the pollen is removed from the baskets and falls to the bottom of the receiver (depending on the design of the pollen trap, this part can be 15–30%, or even 50–60% of the pollen load). A strong colony can harvest 35–40 kg (and even up to 55 kg) of pollen during the season, but a reasonable beekeeper is limited to collecting a relatively small part of the "harvest", no more than 6–10 kg. However, even this amount can significantly improve the economic performance of the enterprise: the market value of 1 kg of pollen is equivalent to the cost of 5–7 kg of honey.

Figure 8.1. A bee returning after collecting pollen placed in baskets on the last pair of legs.

In some countries, pollen load is the main commercial product of beekeeping. For example, beekeepers of the southern regions of the United States in February take bees several thousand kilometres from their home to California to pollinate flowering almonds. The collected pollen load is purchased by pharmaceutical companies and used for the production of dietary supplements and cosmetic products.

For bees, as already mentioned, pollen is practically the only source of the most important nitrogenous compounds, amino acids and peptide proteins consisting of them. Proteins of royal jelly and tissues of larvae and adult bees are all processed, "rebuilt" peptides of flower pollen. In the body of a bee, they are decomposed by enzymes into amino acids, and already from these "bricks", according to a program recorded in the DNA, new peptides are synthesized, naturally with new properties. But plants give bees in the form of pollen not only these "building" materials. In its composition, they receive vitamins necessary for all living organisms (B_1, B_2, B_3, B_6, B_8, B_{12}, C, A, PP, E, D, K, H). Pollen also contains fats, fatty acids and sterols (together they are called lipids), macronutrients (K, Na, Ca, Mg, Cl) and trace elements (P, S, Fe, Cu, Zn, Co, Mo, Se, Cr, Ni, Si). It contains sugars in noticeable quantities (25–35%), but for bees, pollen is only an additional source of these energy or "fuel" elements.

All these components are also contained in beebread; however, doctors have found that biologically active substances of beebread are much easier for the human body to absorb. In addition, pollen, even when dried, is an easily perishable product, while beebread in a dry state can be stored for several years without a catastrophic loss of nutritional properties. And there is one more important difference: in many people, pollen causes a strong allergic reaction, and beebread is almost safe in this regard. Therefore, both pollen and beebread can serve as biologically active additives to compensate for the lack of important components in food. However, beebread, as they say, totally exceeds pollen in its qualities. What is the reason? What kind of magical alchemy is used by bees to obtain such valuable "canned food"?

8.2. No magic, just an ordinary chemical miracle

Now we can say with confidence that, as in the case of honey, the transformation of pollen into beebread begins even during its collection. To bring pollen to the nest without scattering the contents of the baskets, the bee moistens it with saliva. This saliva contains enzymes, some of which are known to scientists and other which perhaps have not yet been discovered. Under the influence of the components of saliva, important changes begin to occur in the pollen. First of all, it loses its ability to germinate, that is, it becomes sterile, incapable of fertilization. The decomposition of complex sugars into simpler ones also begins, and then into the simplest ones, mainly glucose and fructose.

Having delivered the prey to the nest (each basket can contain from 8 to 15 mg of pollen), the foraging bee goes to a place intended for storing beebread. Here she searches for an empty or already partially filled comb cell, lowers the hind legs with a load into it, shakes off the pollen with the help of special spikes on the middle pair of legs and then goes for the next portion of pollen (the cell holds up to 175 mg of pollen, that is, a load brought in by at least a dozen field bees.). Further operations are carried out by hive bees. They crush the pollen, chewing the lumps brought in with their jaws, moistening them with additional saliva and tamping them with their heads.

Again chewing and again this saliva! What, along with it, can get into the product we are discussing besides enzymes? In one of the sources (Kędzia & Hołderna-Kędzia, 2010), it is written that pollen is also moistened by secretions of the pharyngeal glands containing 10-HDA (10-hydroxy-2-decenoic acid), as we recall, by the hydroxy acid of royal jelly with the properties of a strong antibiotic. We did not find 10-HDA in any of the dozens of beebread samples analysed. But in one of them it was possible to find small amounts of unsaturated dicarboxylic acids: 1,10-decene-2-dioic and 1,8-octene-2-dioic. I have suggested that the first of these is the oxidation product of 10-HDA (see section 6.3), while the second can be formed by the oxidation of another component of royal jelly, 8-hydroxy-2-octenoic acid. It can therefore be concluded that the secretions of the pharyngeal glands enter the beebread only sporadically, and that in the process of its "maturation", oxidation of the unsaturated hydroxy acids contained in them occurs. Probably, the appearance of components of royal jelly is associated with the involvement of young nurse bees in pollen processing.

Further transformations of pollen occur under the influence of microorganisms introduced by bees together with saliva, lactic acid bacteria including *Lactobacillus acidophilus* (in nature, lactic acid bacteria are also found on the surface of plants, including on pollen grains). The rapid fermentation of sugars by bacteria is favoured by the rather high humidity of pollen (water content 20–30%) and the optimal temperature, 33–35 °C, in the hive. Under these conditions, the pollen grains swell and their shells are partially destroyed. As a result of fermentation, lactic acid is formed, the content of which can reach 3%. The contents of the honeycomb cells acquire a dark brown colour, an acid reaction (pH 3.5–4.5) and a sweet and sour taste, reminiscent of the taste of rye bread. It is thanks to lactic acid with the properties of a preservative that beebread can be stored for a long time, almost without any loss of its nutritional and medicinal properties. The fermentation process takes about a week. Cells filled to two-thirds of their volume with pollen are closed on top with a thin layer of honey that insulates their contents from the air. As a result, anaerobic (oxygen-free) conditions are created in the cells and this prevents the development of moulds.

Compositional changes are not limited to the formation of lactic acid. During the maturation of beebread, further decomposition of complex sugars occurs (enzymes of saliva, α- and β-amylase "work" to break them down), and the content of proteins and fats reportedly decreases. On the contrary, an increase in the total content of vitamins has been noted, with some new vitamins appearing in beebread. The reason for this phenomenon is not entirely clear. Perhaps they were originally in the pollen in a "bound" state, that is, as part of some kind of complex. It is also known that some microorganisms are capable of synthesizing vitamins. It is reported that beebread the following vitamin content (in mg%): vitamin C (ascorbic acid) – 140–205, B_1 – 0.4–1.5, B_2 – 0.54–1.9, B_6 – 0.5– 0.9, P – 60, A – 50, E – 170 and D – 0.2–0.6. As you can see, beebread is a real storehouse of vitamins, including a complex of vitamins of group B. Their main role in the body is participation in tissue respiration and energy production. These water-soluble vitamins are easily degraded, consumed quickly with increased physical activity and, with the exception of vitamin B_{12} (cyanocobalamin) do not accumulate in the body, so they must be replenished daily.

8.3. Beebread through the prism of chemical analysis

Gas chromatography combined with mass spectrometry (GC-MS), as we remember, is now the most effective method for studying complex (in terms of the number and variety of components) mixtures of organic compounds. In the literature, we will find many hundreds of reports on its application to study the composition of honey, propolis, royal jelly and wax but only in a few publications was the GC-MS method used to study the composition of beebread. It is amazing, inexplicable and, taking into account the above data on the nutritional and health-improving properties of beebread, not a true reflection of it. Beebread deserves more attention!

In one of these studies, the GC-MS method was used to determine the volatile odour components of beebread (Kaškonienė et al., 2008). The authors managed to identify 32 main compounds (three of them, alas, are wrong!), which I will not list.[22] An interesting feature is the presence of sulphur compounds in the emissions: dimethyl sulphide (CH_3SCH_3), dimethyl disulphide ($CH_3S–SCH_3$) and dimethyltrisulphide ($CH_3S–S–SCH_3$). Three sulphides accounted for about 24% of the total volatiles. I have worked with these substances and, please take my word for it, they all have a terrible smell! Many other volatile components of beebread, for example, aldehydes and furans (in total about 18%), do not smell very pleasant. However, in small doses and in certain combinations, these substances give the beebread a pleasant aroma. And this is a common phenomenon: most perfumes with the most exquisite scent contain in small quantities some substances with a completely disgusting smell.

Unfortunately, in the work under discussion, for reasons unknown to me, the authors studied the volatile compounds of only one sample, and that was not pure beebread but its mixture with honey. It is not surprising, therefore, that, among other compounds in volatile emissions, 5-hydroxymethylfurfural (HMF) was found, a substance formed when honey is aged or when it is incorrectly heat-treated.

The second work (Isidorov et al., 2009) used the approach already described in the previous sections, based on the stepwise extraction of the object of analysis with different solvents. As a result of the extraction of five samples of beebread (one each from Russia and Poland, and three from different regions of Latvia), fractions of non-polar (8.2 ± 0.4%), weakly polar (5.4 ± 0.9%) and polar (49.8 ± 5.4%) compounds were determined. All three extracts contained 63.4 ± 5.9% of the beebread weight. This means that about 36% of the mass is the solid residue (insoluble fragments of

---

[22] *In fact, beebread is richer volatile compounds. In the course of the research that we carried out after writing this section, we identified more than 210 substances in six samples of beebread from Poland and Latvia.*

pollen). And this is also not a useless ballast: the fibre contained in it cleanses the gastrointestinal tract and enhances its activity, and thus has a beneficial effect on the digestive organs.

The first fraction of all five samples included 40 substances characteristic of natural beeswax – alkanes, alkenes and esters – with a small admixture of higher alcohols and fatty acids. The second fraction turned out to be richer and consisted of 95 components, the majority of which (64.3 ± 9.0%) were fatty acids. It is noteworthy, and this fact has not previously been noted, that most of all beebread contains α-linolenic acid (ALA). It deserves special attention, and here's why. This is a so-called omega-3 acid (ω-3 acid). Fatty ω-3 acids have a double carbon-carbon bond (–C=C–) in the ω-3 position, that is, at the third carbon atom from the "tail" of the molecule, as can be seen from the example of one of them, eicosapentaenoic acid, EPA (Fig. 8.2).

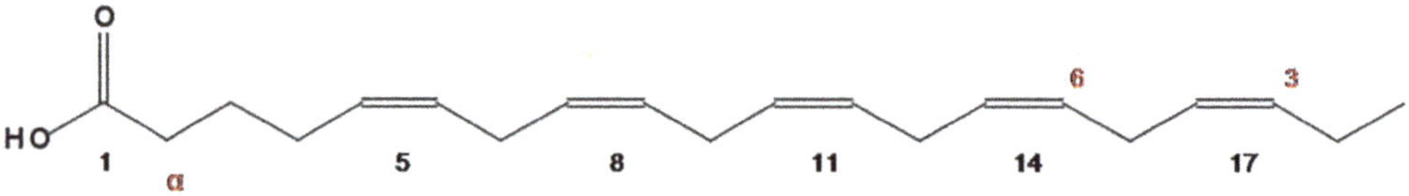

Figure 8.2. Chemical formula of eicosa-5,8,11,14,17-pentaenoic acid

The most important ω-3 polyunsaturated fatty acids are ALA, EPA and docosahexaenoic acid (DHA) containing six double bonds (C=C). Some sources refer to the polyunsaturated acid complex as vitamin F. The health benefits of long-chain ω-3 fatty acids are well known. Suffice to say that their content is approximately 3% of the dry weight of the grey matter of the brain (EPA is the main one). There are also many of them in the photoreceptor cells of the retina. Therefore, ω-3 acids are vital for the functioning of the brain and eyes. They are also necessary for the proper functioning of the cardiovascular system. Their deficiency in a child's diet can seriously affect the child's development. The human body is not able to synthesize these fatty acids from simple substances and therefore they must enter it directly from food. The food richest in ω-3 acids is sea fish.

Studies conducted back in the 1970s showed that the indigenous people of Greenland, who consumed a large amount of oily fish, suffered from hardly any cardiovascular diseases and did not have atherosclerotic lesions. Other indicators, such as blood pressure and "bad cholesterol", that is, harmful low density lipoprotein (LDL) and very low density lipoprotein (VLDL), were significantly better than those in other populations.

It would seem that it is simple: you need to give a child as much oily sea fish as possible, and everything will be fine. And for the elderly, it is useful for preventing the occurrence of atherosclerosis. However, saltwater fish, especially oily fish, contains such large amounts of mercury that the Food and Agriculture Organization of the United Nations (FAO) discourages even adults from eating it too often. What can we say about children?

But not everything is so hopeless, my Reader: subsequent studies have shown that the most important ω-3 acids (EPA and DHA) can be formed in the human body from the shorter-chain ALA, the very one, which is found so abundantly in beebread! It accounts for 29–37% of the ester fraction in beebread which has a low content of ω-6 acids, which can inhibit this synthesis. Beebread contains other lipids that can reduce the content of "bad cholesterol". These are the so-called phytosterols, primarily β-sitosterol which can reduce the absorption of cholesterol in the intestines.

Well, what about phenolic compounds, which, as we remember, have antioxidant properties? In all five samples of beebread, we found small amounts of *p*-coumaric acid (367±101 μg/g), and other cinnamic acids, ferulic and caffeic, were present only in trace quantities. In somewhat larger quantities, beebread contained two flavonoids, kaempferol (492±350 μg/g) and isorhamnetin (1086±720 μg/g). Among other flavonoids, which are contained in most samples in trace quantities, I will name chrysin, apigenin, naringenin and quercetin. In order of magnitude, our data on the amount of flavonoids are consistent with those given by the Lithuanian authors (3141±935 μg/g) for the three beebread samples studied by them (Čeksterytė et al., 2006). They discovered the largest amounts of flavonoids in two samples of beebread with a high winter rape pollen content. There was less of them in beebread with a high content of white and red clover pollen.

In another work, Lithuanian authors compared the antioxidant capacity of mono- and polyfloral honey and beebread (Baltrušaitytė et al., 2007). Again, we have to express regret that out of nine samples, only one was a beebread extracted from combs, and the rest were a mixture with honey (1:1) or with honey and wax. Nevertheless, the authors found that extracts from beebread are much more active in sweeping away free radicals than honey. This activity, in my opinion, correlates weakly with the content of polyphenol compounds, both discovered by us and by the Lithuanian authors. This may be due to the fact that some of the polyphenols in beebread could not be discovered due to the fact that they are part of more complex substances, for example, in the form of glycosides (that is, compounds formed by sugars and non-carbohydrate fragments, including flavonoids). Such compounds are very widespread in the plant world.

Indeed, the methanol extract, in which 92 compounds were registered, contained glycosides (Isidorov et al., 2009). Unfortunately, at that time we were unable to identify most of them. The main components of methanol extracts are simple sugars (glucose, fructose and mannose), disaccharides and sugar alcohols (mannitol, sorbitol and others), as well as sugar acids. The content of phosphorus compounds, phosphoric acid and its esters is also noticeable.

And, in conclusion, there is one more (second in a row) riddle related to free amino acids. I really hope that the Reader remembers the first one concerning the presence (or rather, absence) of these components in royal jelly. What has already been forgotten? Then I beg you to look again at section 3.5. But back to the beebread. One of its positive qualities is the presence of free amino acids. In four out of five samples of beebread, we did not even find traces of these very compounds. But they were in the fifth test! Mysticism, and nothing more.

For those who were too lazy to fulfil my request to look at section 3.5 again, let me remind you: free amino acids in royal jelly are an artifact (erroneous result) associated with the preparation of samples for analysis. As soon as the larva is injured, amino acids "polluting" it appear in the royal jelly. In our opinion (Isidorov et al., 2009), even in the case of beebread, it's all about sample preparation, with the only difference that the product is not "contaminated", but amino acids disappear from it! The fact is that I was handed the fifth sample by the well-known Latvian beekeeper J. Sulutaurs immediately after he extracted (literally, knocked out) beebread from previously frozen combs. It is this fresh beebread that athletes willingly buy from him. All the other samples we examined were dried to prevent mould. Dried beebread can be stored for a very long time at room temperature; it can be sent to the customer by mail. Fresh is stored only frozen.

What happens to the beebread during drying? For an organic chemist who has read the above information about the composition of beebread, the answer is obvious: the Maillard reaction takes place. It is a chemical interaction, accelerated by heat, between amino acids and reducing sugars.

The amino acid interacts with the carbonyl group (–CH=O) of the sugar, with the release of a water molecule. In principle, this process is reversible; however, the reaction product (glucosylimine) quickly rearranges into ketosamine, called the Amadori compound. That's it, the amino acid has disappeared! Along with it, alas, a piece of the usefulness of beebread also disappeared.

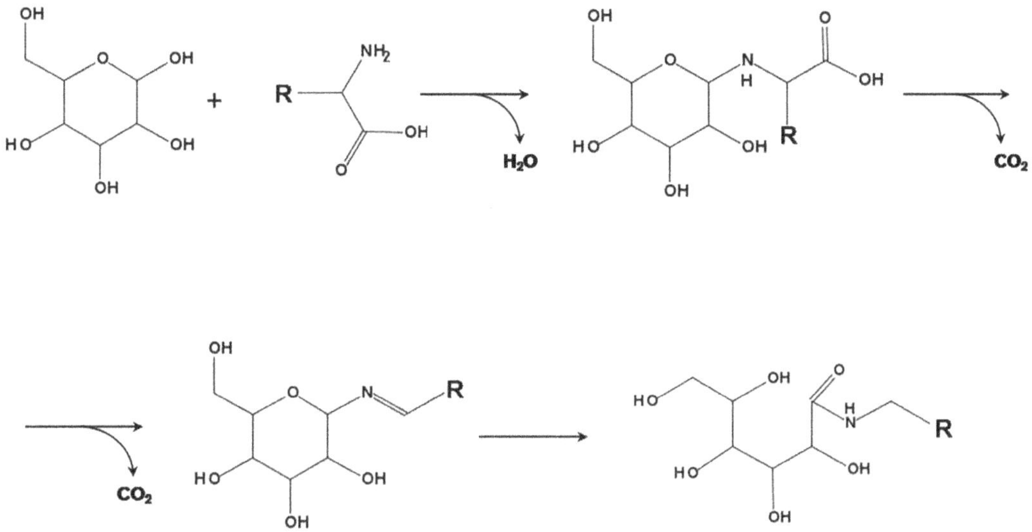

Figure 8.3. Schematic representation of the reaction of a monosaccharide (glucose) and an amino acid and the subsequent formation of the Amadori product

What practical recommendations can be drawn from the above? It depends on who the recommendations are addressed to. For an ordinary consumer, they are very simple: try to buy fresh (not dried even in mild conditions!) beebread, but store it in the freezer. A small portion of beebread taken out of the freezer in the evening will thaw out by the morning and will be ready for use.

For those young scientists who, as I hope, will pay attention to beebread and undertake the work of studying its composition, I would recommend that they approach the "history" of the product very carefully: how, where and when it was obtained and what processing it was subjected to. And it is worth working with beebread, since it opens up a vast field of activity for the scientist, not least due to the fact that beebread is in most cases non-monofloral. Consequently, wide variations in its pollen and chemical composition can be expected. It is also important that this area of scientific research is hardly "trampled" yet. Indeed, the number of publications on the composition of beebread can be counted on one hand, and the number of samples studied so far is ridiculously small. The use of geo-botanical and biomedical data and the use of modern methods of processing research results (primarily, chemometrics) will bring unheard-of satisfaction and serve as the basis for dissertations of any level. So, I wish you success!

CHAPTER 9
Stay healthy, bees!

9.1. About the health and treatment of bees in general terms

Year on year, there is growing concern due to the deteriorating health of honey bees, like all other insects belonging to the Anthophila section and the Apidae family, which includes not only domesticated, but also wild solitary bees and bumblebees (about 170 genera and more than 5000 species). Over the past 100 years, 50% of wild bees from the Eurasian and North American continents have disappeared from their historical areas. It is reported that the abundance of four native species of bumblebees has decreased by 96% over the past 20 years, and three species are considered extinct. This worries not only beekeepers but also the general public, since the problem is constantly in the spotlight of the media. The latter, unfortunately, often cannot resist the temptation to alarm readers, listeners and spectators with, for example, the constant repetition of the absurd assertion attributed to Albert Einstein about the imminent and inevitable death of the human race after the last bee disappears on Earth. Einstein, although he was a theoretical physicist and did not specifically study biology, was not so ignorant as to not know that cereals, potatoes and many other root crops do not need pollinators. So starvation does not even theoretically threaten people. However, the truth is that without bees, our planet would not be a blooming garden since 80% of wild plants depend on pollinating insects*[23]. People would have to forget the taste of almonds, cherries, apples and even onions (the dependence of these crops on pollinators is 90% or more).

The reasons for the bees' plight are diverse. We have to admit, whether we want to or not, that the main cause is the ever-mounting pressure on nature, on the biosphere of the Earth, by an ever-growing human population with its ever-increasing needs, both vital and imaginary. Due to this pressure, environmental conditions deteriorate, which often leads to the death of populations mainly due to various diseases. An additional factor aggravating the situation is the undesirable and

---

[23] *Recently published results of research by German scientists showed that over 27 years (1989–2016), the biomass of flying insects in 97 protected (!) territories of Germany decreased by 76% (Hallmann et al., 2017).*

practically uncontrolled migration of pathogens over long distances. With human participation, of course.

According to the unanimous opinion of Polish beekeepers, with whom I had a chance to discuss the health of bees, in the current conditions bees of the *Apis mellifera* species have no chance of surviving without constant medical help from humans. Until relatively recently (some 70–75 years ago), a family that "ran away" and settled in the forest could live and share quietly, giving rise to new families. But this is in the past. Nowadays, if some family rises and leaves the apiary "from a low start" to settle in the forest, where no one will take away any of their honey, bee bread or propolis, the probability the family's death is great. Such a family that escaped from an apiary and runs wild will inevitably bring diseases with it and it will not be easy for the family to survive without medical assistance. What this pathogenic burden is, the reader will understand from what follows.

Some amateur beekeepers from different countries, including Poland, do not agree with this opinion. They believe that the «domestic» bees will cope with all the problems if they cease to interfere with constant examinations, if they will cease to completely select the stocks they have harvested, replacing them with low-value artificial feeding and, most importantly, if they will stop «treating» them with the help of synthetic chemicals. These beekeepers follow the guidelines set out and form self-help groups (e.g. Fort Knox Association; http://bees-fortknox.pl) based on the principle of guaranteeing mutual compensation for possible losses of bee colonies intended for natural selection.

With all my heart, I wish success to those who have embarked on the difficult path of «natural selection» of disease-resistant bees!

The list of bee diseases known to veterinary science includes many tens of items. Some of them are identified by the legislative norms of the European Union by two special categories: diseases that are subject to mandatory registration, and diseases that must be mandatorily eradicated. In Poland, the first category includes five items: European foulbrood; varroosis; and acariosis, as well as infestations of the *Tropilaelaps* mite and the small hive beetle *Aethina tumida*. The second category is more severe, but in the case of bees, it carries only one disease, the American foulbrood.

In short, domesticated bees need treatment. What about drugs? I must say that the observed increase in the incidence of bees is not accompanied in the European Union countries by an increase in the supply of medicines, rather the opposite. There was a time when antibiotics and sulphanilamides could be used to treat bees' bacteriosis, as well as fumagillin in the treatment of fungal infections, but this is currently prohibited by law due to a well-founded fear that they can pass into the honey and other beekeeping products (other negative consequences of antibiotic use will be discussed later). The antibiotics oxytetracycline and rifampicin, as well as the sulphanilamide drug norsulphazole and fumagillin-B are approved for use in Russia, but this is one of the reasons why Russian honey cannot break into the European market.

As a result of these prohibitive restrictions, beekeepers in the countries of the European Union now do not have any synthetic medicines for the treatment of bacterial and fungal (as, indeed, viral) diseases of bees. In the case of combating parasites, the situation is not so hopeless. For example, in Poland, two acaricides were allowed to circulate: fluvalinate, a synthetic insecticide from the third-generation of pyrethroids, and amitraz from the amidine group. The latter is considered "less toxic to mammals"; however, many beekeepers are wary of it, since vomiting and bradycardia are common symptoms in those working with it. Moreover, in the literature reported (Ellenhorn et al., 1997) about 41 cases of deadly amitraz intoxications have been reported.

It is not surprising, therefore, that beekeepers are independently looking for ways to treat bees, mainly by resorting to herbal remedies. On the Internet you can find many posts and discussions on this "sore" subject. They also follow with great interest breeders' experiments in their efforts to develop lines of bees that are resistant to disease, especially to varroatosis. The subsequent sections of this chapter are devoted to the diseases of bees, their treatment and prospects in this direction.

9.2. The most endangered species

The honey bee (*Apis mellifera* L.) is, in the words of the Polish science fiction writer and philosopher Stanisław Lem (1921–2006), one of the species that are most at risk (Lem, 1989). According to Lem, it is in "... species whose populations are most numerous and that live so crowded that panmixia[24] takes place in the field of sexual relations". In such a situation, the likelihood of lentivirus[25], which have the maximum lethal force, increases. Bees, wasps, ants and termites – socially living insects – already at the long stages of evolution have protected themselves from this danger, excluding panmixia from their lives (unlike representatives of another social species, *Homo sapiens*). Nevertheless, all these social insects are at very high risk of bacterial infection.

To survive in crowded conditions in nests in which temperature and humidity are at a level optimal for the development of microorganisms, bees had to develop a very effective defence strategy against infectious diseases during their evolution. The first element of such a defence is, without a doubt, the relentless concern for nest cleanliness (Visscher, 1980). The habit of removing dead material and other impurities has developed in all social insects: bumblebees, termites and ants. Interestingly, in ant nests there are special burial chambers, and "funerals" are carried out by a group of specialized individuals who have limited contact with other family members (Visscher, 1983).

In addition to this "hygienic" behaviour, bees constantly destroy infected larvae. Elements of adaptive behaviour also include the flight of infected bees during wintering, when "... at the low cost of the death of separate individuals, the "disease load" in the entire superorganism decreases ..." (Kasperek & Paleolog, 2011). It is clear that all these measures are necessary, but are not sufficient to maintain the health of a family of tens of thousands of members.

Another element of the survival strategy could be a high level of individual defence of the organism acquired during evolution, including not only anatomical and physiological barriers, but also other, cellular and humoral defence reactions. These can be seen in the way that, in response to the danger of an infectious disease, certain processes responsible for immunity begin to act within the organism. This

---

[24] *Free crossbreeding of heterosexual individuals in a population.*

[25] *Viruses with a long phase of latent development. Lentiviruses include the HTLV-III virus, which causes HIV infection.*

happens through the "inclusion" (biologists use the term "expression") of special genes that control the synthesis of antibacterial peptides (AMPs), or are involved in cellular immunity. Such a mechanism acts in the body of all animals, including honey bees. The haemolymph of a bee after the penetration of a pathogen (or after an insect injury) acquires antimicrobial activity due to the development of a number of AMPs: apidaecins, abaecin, hymenoptacin. However, recent studies (Evans et al., 2006) have established a paradoxical fact: the level of this kind of individual (that is, internal) physiological protection in bees is lower than in non-social insects! Consequently, in the course of the evolution of honey bees, some other, and no less effective protection mechanism (Evans & Spivak, 2010) inevitably should have developed, with some advantages compared to individual protection.[26] This is the so-called mechanism of group immunity, which will be discussed below.

Let us recall once again that bees do not live in a sterile environment, and protection from pathogens in conditions of extreme crowding is a matter of life and death for them! During the collection of nectar and pollen, they circulate over large areas and come into contact with honey plants populated by various microflora. A study by Russian scientists on the enteroflora of linden blossoms, white clover, medicinal dandelion, raspberries and fireweed, all of which are the main honey plants, showed a high degree of population with many species of bacteria of the genera *Hafnia*, *Klebsiella*, *Citrobacter*, *Enterobacter* and *Erwinia* (Rechkin & Evteeva, 2007; Evteeva et al., 2010).

Together with nectar and pollen, as well as on their body, bees constantly (and, presumably, in considerable quantities) bring these microbes to the hive. Among them, the causative agent of aspergillosis (stonebrood), *Aspergillus flavus*, enterobacteria, causing hafniosis (*Hafnia alvei*), *Streptococcus pluton*, *Enterococcus faecalis* and *Streptococcus faecalis* cause damage to bees with European foulbrood (the last two types of bacteria come from polluted water sources). In addition to this, bees are constantly threatened by an invasion of various kinds of parasite insects and other pests. Nevertheless, for many hundreds of thousands of years, the bees have somehow managed to cope with all these misfortunes. The changes in the environment that took place during this time, such as, for example, climate

[26] *There is strong evidence that this already relatively weak protective barrier, if it does not break down, is significantly damaged due to the increasing intake of various agricultural preparations, fungicides and pesticides, as well as the chemicals used to treat bees. All this cannot but lead to a weakening of individual immunity.*

fluctuations and related changes in the species composition of vegetation, were so slow that the bees managed to adapt to them. However, starting from the second half of the nineteenth century, and especially in the twentieth century, these processes throughout the world, and particularly in the Northern Hemisphere, have become threatening. Deforestation, landscape change, intensive agriculture with the inevitable use of artificial fertilizers and chemical plant protection products, catastrophic pollution of air, soil and surface water by chemical compounds alien to nature (called xenobiotics), and the emergence of new, genetically modified species of organisms – this is not an exhaustive list of these changes. How do bees feel against this background? Unfortunately, the answer is: bad!

In recent years, the health of bee colonies has catastrophically deteriorated. Suffice it to recall the Colony Collapse Disorder (CCD), which manifests itself in a strong decrease in worker bees, and then in the death of a sick family. Against the background of a decrease in the number of bees, it was noticed that in the hives there remains a queen bee with a small number of workers, as well as unused supplies of bee bread and honey. The reason for this phenomenon has not yet been elucidated, but there is a long list of factors that are likely to affect the occurrence of CCD in bees. These include the invasion of the external parasitic mite *Varroa destructor*, a decrease in bee immunity, nosematosis caused by a protozoa related to fungi *Nosema apis* and *Nosema ceranae*, Israeli Acute Paralysis Virus (IAPV), pesticide poisoning (mainly from the neonicotinoid group), and environmental changes caused by global warming.

In my opinion (as well as in the opinion of a number of other scientists), the main cause of this syndrome is the spread of varroatosis, while other biological factors (decreased immunity, nosematosis, IAPV and other viruses) can be derivatives of varroatosis. It is known that female mites parasitize on bees affected by them, and the consequence is a decrease in the survivability of bees and a sharp decrease in resistance to infection. The Varroa mite has been found to carry many bee pathogens, such as Chronic Bee Paralysis Virus (CBPV), Acute Bee Paralysis Virus (ABPV) and Slow Bee Paralysis Virus (SBPV). Mites infected with the Deformed Wing Virus (DWV) are able to successfully transfer it to pupae, most often at the "white eye" stage. It is this virus that is responsible for the appearance of crippled bees in bee colonies that cannot fly.

Beekeepers faced mass deaths of bee colonies even much earlier. A little over 100 years ago, in 1906, the beekeepers of the Isle of Wight, located off the southern coast of England, sounded the alarm. For unknown reasons, most bee families died out almost simultaneously, with the appearance of creeping, «trembling» and unable to fly bees. This phenomenon is called the «Isle of Wight Disease». It took a long time to find out the nature of this disaster. Scientists came to the conclusion that it was caused by a number of factors, such as infection with a previously unknown chronic paralysis virus (SBPV), bad weather, and too many bee colonies in a limited area. The second wave of the mass death of bees in England occurred in the 1950s. And in this case, it was caused not by one, but by several reasons.

It is more correct to call such phenomena as CCD and the "Isle of Wight Disease" not a disease, but a "syndrome", which in Greek means "managed together". In each syndrome, the main link can be distinguished; however, the development of the pathology depends to a large extent on its other components. Often, the sub-lethal effect of one link of the syndrome increases the mortality of the other. This can happen, for example, with the combined effect of parasites and pesticides on bees.

As we can see, bees have a powerful protective potential that has allowed them to survive all sorts of adverse factors that have repeatedly arisen on Earth. However, this potential turned out to be too weak in our time. Endless and painful questions arise. Who is the begetter, and what to do about it? Unfortunately, the first of these can be answered quite confidently: man is the author of everything! The history and course of the development of varroatosis, a parasitic disease that affects both bee brood and adult bees, convinces us of this.

9.3. Varroatosis

9.3.1. Anamnesis and current condition of the patient

The *Varroa destructor* mite was discovered in various Asian countries at the beginning of the last century in the families of the eastern bee *Apis cerana*, but it did not do much harm to them: after the numerous wanderings of *V. destructor* through the new generations of *A. cerana*, a "dynamic equilibrium" between these species occurred[27].

The defence mechanism developed in the course of evolution (together with *Varroa*) is based in eastern bees, firstly, on an extremely strong instinct for self-cleaning of parasites and, secondly, on the skill of removing them from the brood cells. In addition, *A. cerana* bees have the ability to recognize cells in which drone larvae (*Varroa* breeding in the families of these bees occurs exclusively on drone larvae) are severely affected. Adult bees leave such cells sealed, and since weakened drones cannot get out of them on their own, they die and remain buried in them along with the parasites. This significantly reduces the level of *Varroa* infestation in the bee colony. Unfortunately, the honey bee *A. mellifera* does not have the skill and ability to deal with the parasite characteristics of the eastern bee. Moreover, by allowing the *Varroa* mite to move into the colonies of the eastern bee, people ensured it a huge success in life, since it had the opportunity to reproduce not only on drone brood but also on a much larger brood of worker bees. This leads to an exponential increase in the number of parasites in the family. Therefore, in families that have not undergone systematic treatment, the number of parasites is constantly increasing, which ultimately (usually no later than the third year) leads to their death. According to the Polish scientist professor Wojke (Woyke, 2001) "the tragedy began only when the parasite *Varroa* switched to the honey bee *Apis mellifera*", which people brought to Asia because, under favourable conditions, the *A. mellifera* bee is much more productive than *A. cerana*. What followed next is simple: the import of packages (layering) or even entire families of bees from Asia to European countries led to the infection of our bees with *V. destructor* mites. While in the opposite direction the causative agents of diseases that were previously unknown in the East were introduced: bacteria that cause European Foulbrood (EFB) and the Sacbrood Virus (SBV). With bitter irony, we can say that no one was left out ...

9.3.2. "Case history" and the patient's condition

---

[27] *As a result of intensive research, it was found that several representatives of the Varroidae family are found in the families of A. cerana. Besides V. destructor, other species include V. jacobsoni, V. unterwoodii and V. rindereri. However, these species, fortunately, cannot reproduce on the brood of A. mellifera. In addition, it was found that there are two varieties (haplotypes) of V. destructor: "Korean" and "Japanese-Thai". The first, more dangerous, spread to the continents of Eurasia, Africa and North America, while the second is found (in addition to the countries whose name it bears) in South America, not causing such losses as the first. Therefore, later on, under the name "Varroa", we will understand the "Korean" type of parasite.*

Varroatosis entered Europe relatively recently. In the USSR, as early as 1978, a document order "On urgent measures to combat varroatosis in 1978" was produced in connection with the manifestation of this disease. In Poland, the disease was first registered a little later, in 1980. However, it only took five years for the disease to spread throughout the country, and now it is the cause of the greatest losses in beekeeping. And not only in Russia and Poland. The main reasons for the rapid spread of varroatosis in Europe are migratory beekeeping and the sale of infected splits. The spread also favoured by the appearance of "wandering" bees, and robbery, which is especially often observed in late summer and autumn. To this can be added the installation of honeycombs with an infected brood in the hives in order to balance the forces of different families in early spring and autumn in preparation for winter. In other words, there are a lot of paths to spread the varroatosis infection, and, unfortunately, we must admit that they are trodden mainly by beekeepers themselves.

Varroatosis differs from other known diseases of bees in that it affects both brood and adults; moreover, the *Varroa* mite harms the bee family all year round, unlike most other diseases that are seasonal in nature. In late autumn and winter, when there is no brood in the colony, its reproduction stops. At this time, all the mites are on the bees, and during the winter, some of them, but not more than 10%, die. Parasites hibernating on bees cause them anxiety, which prevents the formation of a normal club, which loosens up. As a result, it is difficult for bees to maintain an optimal temperature, as they are numb from the cold and often crumble.

The life cycle of the *Varroa* mite consists of two separate phases: the reproductive phase and the so-called "phoretic" phase. The first of them passes in the cells of the bee brood, and during the second, the parasites move to adult bees, which serve mainly as a "vehicle" that allows *Varroa* to spread in the environment. Robber and stray bees, as well as drones that fly into other hives, are involved in the infection of bees in other families and in other apiaries.

"Phoretic" individuals parasitize on adult bees, while larvae at various stages of their development serve as a food source for mites in brood cells. During bee hatching, offspring of the female *Varroa* enter the cells with larvae and (after their sealing) lay eggs, one unfertilized and several fertilized. A male develops from an unfertilized egg, and females develop from fertilized eggs. The male mates with his sisters, after which he dies, and the newly fertilized females leave the cells along with the young bee. The parasite larvae go through protonymphal and deutonymphal stages of development to adult imago, consistently receiving food from the body of the bee larva. Interestingly, *Varroa* females, who are well orientated in a bee nest, possess excellent smell, and perfectly feel the temperature, feel best at a temperature no higher than 30 °C. Therefore, most often they prefer to populate cells in the lower parts of the frames, where the temperature is slightly lower than in the centre of the nest. Moreover, the fecundity of the parasite is associated with climate. In hot climates, more female mites are born that are not fertile.

The damage caused by parasites depends on their number in the family. If only one female enters the cell with the larva, then, as a rule, later, a weakened but living bee is born whose life expectancy is less than normal. However, with an increase in the number of mites in the nest (with an exponential increase in the number of the parasites, their number can increase 115 times in 100 days!), 2–3 females can enter one cell with bee brood. Bees with various kinds of defects, often incompatible with life, gnaw out from such cells. If an even larger number of *Varroa* females enter the cell (the largest recorded number of parasite individuals in one cell was 18!), then a living worker bee never appears from it. In late autumn and winter, when there is no brood in the nest, mite reproduction ceases. But, even if it does not reach the death of a bee, the damage to its health is very great. In addition to reducing the bee's lifespan, it is expressed in a reduction of up to 25% of its body weight. There is also a decrease in the ability to orientate and learn. In the case of drones, a decrease in sperm count is observed (Chorbiński, 2012).

The larvae of the parasite pass through various stages of development up to the adult imago, constantly receiving food from the brood of the bee, while the adult, "phoretic" females, parasitize on the adult bees infected with them. It seems that, until very recently, not only ordinary beekeepers-practitioners but also scientists, whose main object of attention is the relationship between *Varroa* and honey bees, were deeply mistaken in the seemingly obvious question: what does the mite eat?

All available sources, including the most solid scientific research, say that hemolymph, which circulates in the body of larvae and adult bees, serves as nutrition. The basis for this statement was, on the one hand, the structure of the oral apparatus of females of the piercing-sucking type, and, on the other hand, the analogy with it of the mouth apparatus of insects sucking the blood of mammals. Recently, however, a group of American researchers undertook an extensive study of this problem and came to the conclusion of a completely different type of nutrition (Ramsey et al., 2019).

A prerequisite for the revision of the established view was a comparison of the nutritional value of mammalian blood and insect hemolymph. So, if the blood is about 40% composed of nutrients, there is less than 3% in the hemolymph of the bee. The rest is water. It is highly doubtful that the parasite will survive on such a poor "syrup", even if it is completely drained (for a worker bee, the amount of hemolymph is only a few milligrams).

On the other hand, the composition of *Varroa* waste indicates that their diet consists of matter containing a large amount of fat and protein, but not enough water. Scientists came to the conclusion that the bee's fatty body, and not hemolymph, is the nutrient material for the parasite.

The fatty body is loose cell tissue lining the internal organs and internal walls of the body of the larva and adult bee. According to the functions performed, the fatty body is considered an analogue of the liver of mammals. The fatty body reaches its greatest development in larvae that pass into the stage of the pre-pupae. In adults, it is most developed in bees preparing for wintering. The main function of the fatty body is the synthesis, accumulation and reverse transport of reserve carbohydrates, fats and proteins. In the cells of the fatty body (trophocytes), fats accumulate in the form of separate drops, carbohydrates in the form of glycogen (glucose polymer), and proteins in the form of granules 1–3 microns in size.

How does *Varroa* get food from the fatty body? Its digestive system, resembling a tube, does not contain digestive enzymes, and the method of its nutrition can be called "external". Parasites inject saliva containing digestive enzymes into the body of a larva or adult bee, and then the resulting liquid homogenate is sucked out. This method of feeding is characteristic of predatory ticks, which are very widespread in nature.

What gives us new knowledge about how the formidable bee parasite feeds (to which we owe American scientists)? Two aspects can be distinguished here. The first is related to the difficulty of conducting experimental studies using live female mites, which do not want to live long and lay eggs if they are fed with hemolymph. However, life expectancy (and female fertility) doubled when they were fed with a fatty body (Fig. 9.1).

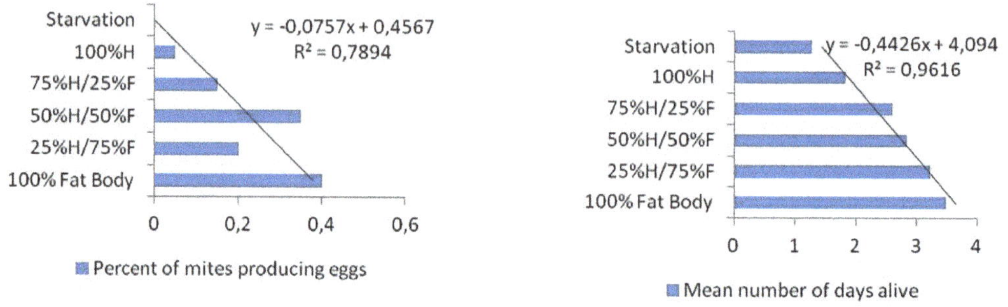

Figure 9.1. Dependence of the average life span of V. destructor females under laboratory conditions and their fertility on the composition of the feed given to them. F – fat body, H – hemolymph (Ramsey et al. 2019).

The second aspect is related to the possibility of treating the disease. It is possible that failures in the development of medicinal preparations against varroatosis, given to bees with food, are in no small measure related to the erroneous idea of the way the parasite feeds. These drugs were prepared with an eye to their circulation in the hemolymph, that is, they had to have hydrophilic properties. Meanwhile, in the fatty body, bees can accumulate only lipophilic substances, that is, easily soluble in fats, to a concentration level toxic to the parasite.

9.3.3. Current approaches to combat varroatosis

The history and picture of the disease is clear, but what is the current state of the disease and the fight against it?

At the beginning of the new millennium, the spread of the disease became catastrophic. This is evidenced by reports from different countries, according to which up to 70% of families died due to varroatosis, with an acceptable level of 10–15% (Neuman & Carreck, 2010). It is not surprising, therefore, that already in the early 1980s, the foundation was laid for research aimed at combating this parasite and finding ways to treat families with artificial chemicals, acaricides, at the forefront. To combat *V. destructor*, formamidines (amitraz, cymiazole) are used, as well as pyrethroids supposedly less dangerous for bees, most often acaricide flumethrin. However, in recent years, an increase in the immunity of the parasite to this acaricide has been noted. The dubious "merit" of such selecting (unplanned, of course) of fluvinate-resistant mites probably belongs to Sicilian beekeepers

(Eischen, 1995). From Sicily, it began his wanderings to the north with migratory ("nomad") apiaries. When a wave of parasites reached the north of Italy, the use of this acaricide did not bring the desired effect anymore because on the way they acquired undesirable resistance to it.

Accelerated selection of the parasite in the direction of immunity to therapeutic drugs most likely could be associated with their improper use. For example, reducing the time spent in the hive with acaricide-saturated strips, or conversely, an excess of exposure time compared to a recommended manufacturer. In such cases, only the most susceptible individuals of *Varroa* are killed, while the most resistant survive and pass on this trait to their offspring. The same thing happens when using medical materials that have expired or that have been stored under inappropriate conditions, such as being exposed to sunlight or high temperatures.

In Poland and other European countries, the use of veterinary medicinal products is legally controlled. To combat varroatosis, four drugs have been officially approved: Apiwarol AS in the form of amitraz-containing fumigation tablets, strips Biowar 500 with the same acaricide (the producer of both products is Biowet, Puławy), Bayvarol in the form of flumethrin-saturated strips (Bayer, Germany), and Apiquard (VitaEurope, England). The last drug is a fairly volatile gel containing thymol. This is the only commercial preparation approved for use in the so-called "ecological apiaries", since thymol is found in the nectar of some honey plants and, therefore, in nectar honeys.

According to Professor Lipiński et al. (2011), who conducted a summer study of *V. destructor* populations in 31 apiaries in Poland, in 17 of them (i.e. 55% of apiaries), the mites were completely immune to tau-fluvalinate, and in five others (16%) partially refractory. The dynamics of this phenomenon does not inspire optimism, since it indicates a rapid increase in *Varroa* immunity against pyrethroids: in 2007 only isolated cases of partial parasite immunity were recorded. In addition, with the use of all the mentioned preparations, there is a danger of bee products – especially honey and wax – being polluted.

"Ecologically safe" in the fight against varroatosis is the use of lower aliphatic acids: formic and oxalic. For this purpose, interframe space is often sprayed with aqueous solutions of these acids. However, the increase in humidity associated with the use of water solutions can trigger the development of fungal infections, such as ascospherosis (chalk brood), caused by *Ascosphaera apis*. Spores of this fungus are extremely resistant to most disinfectants used in practice and can remain viable for 2–4 years. Therefore, beekeepers are increasingly resorting to other methods of using acids: treating hives in pairs, in particular by sublimating oxalic acid at elevated temperatures.

Formic acid evaporating from the dispenser (usually a 65% aqueous solution is used) has a destructive effect on the respiratory apparatus of the mite, which leads to its death. It is the only antivarroatous agent capable of penetrating, to some extent, sealed cells and killing the parasites in them. As a rule, it is recommended to treat families twice with this remedy. The first time this is done after the main honey collection, and the second time this is done at the end of September or at the beginning of October, depending on the weather conditions. However, one should remember its toxic effect on bees and larvae, which can be observed at an air temperature above 25 °C and with poor ventilation of the hive (on the other hand, the air temperature should not fall below 12 °C, since the acid volatility greatly decreases and the treatment becomes ineffective). In addition, during the use of formic acid, a number of negative phenomena can be observed, such as strong anxiety of the bees (which can begin to destroy the brood) and the termination of the laying of eggs by the queen. In extreme cases, it can lead to the death of the queen, being suffocated in the club formed by bees. Oxalic acid vapours can also infect the respiratory organs of bees, and beekeepers too.

Unfortunately, it turned out that the *V. destructor* gradually becomes resistant even to these substances! According to Russian beekeepers, this happens after 4–6 years of using oxalic acid, and in some cases, its effectiveness decreases after 2 years (Bogomolov & Yarankin, 2011). The same applies to formic acid, which, in addition to everything, causes a terrible bitterness in the bees, and this is a factor negatively affecting their performance and vitality.

The above is directly related to the second of our questions: what to do? Effective chemical preparations for combating bee diseases (and not only varroatosis) have been largely discredited. This is due to various reasons, whether it is a loss of effectiveness due to the development of resistance in parasites or whether it is high toxicity, both for bees and for humans. An example of such toxic substances is amitraz, which (for lack of anything better) continues to be used in many countries, despite the widespread increase in resistance to it among *Varroa*.

The exceptional severity of the problems described above has forced both beekeepers and scientists to search for other alternative chemotherapy methods to combat varroatosis, and "independent" infectious bee diseases accompanying it. In this sense, several promising areas are outlined:

- use of the "defence potential" of the bees themselves,
- zoo technical and biological methods of combating diseases,
- use of natural herbal preparations for the treatment of bees,
- use of non-drug methods of combating varroatosis.

It is possible that the severity of the health problems of bees will also force beekeepers to change their farming methods, particularly to abandon nomadic beekeeping, or to strictly limit it legally. I have repeatedly said that migrations contribute to the spread of varroatosis, as well as many other dangerous diseases of bees, such as, for example, American foulbrood.

In a recent scientific publication (Parish et al., Apidologie, 2019), attention was paid to an important aspect related to apiary migration. Bees in the aforementioned publication are considered as a potential peddler (vector) of infections. Using as an example two microscopic fungi, Botrytis cinerea (a causative agent of grey rot) and Colletotrichum acutatum (a causative agent of anthracnose, putrefactive strawberry disease and leaf necrosis), the authors studied the survival of spores that passed through the digestive tract of bees. As known, adult bees defecate outside the hive at a maximum distance of 5–6 km from it. However, with nomadic beekeeping, they often turn out to be transported in less than a day for many hundreds and even thousands of kilometres. The authors of the cited work found that the survival of the spores of the fungi studied by them, isolated from faeces, is at the level of 1%. It would seem that with such a low survival rate there are no problems; however, in every hive, many tens of thousands of worker bees travel (and in the whole migrating apiary there are many millions of them!) and this determines the high probability of the transfer of live spores in quantities that can cause a sudden outbreak of infection. This implies the need for quarantine restrictions on the transportation of bees from areas in which economically important fungal infections are recorded.

The existence of their own "defence potential" in bees is evidenced by the fact that on the walls and on the frame in the hive it is not possible to detect an excess of microorganisms. This can only be explained by the fact that microbes brought to the hive along with nectar and pollen almost immediately die in it. The main "culprit" of this is propolis. It can be assumed that the substances contained in it, and not the

AMP hemolymph, protect bees from these microorganisms. This idea is suggested not only by the relatively low level of individual protection, but also by the fact that many of the bacterial species mentioned above have developed mechanisms of resistance against AMPs (Żyłowska et al., 2011), but not to propolis.

With a high degree of probability, it can be assumed that the main element of social (i.e., external) immunity, which guarantees defence against morbidity at the level of a whole family, is the use by bees of plant balsams, which, when mixed with wax, we call propolis. In other words, the accumulation of propolis in the hive and its further use for disinfection, especially cells for growing brood, plays an extremely important role in the life of bees. It has long been noticed that bees that settle in the favourite hollow of a tree, first of all begin to "scrape" the rotten wood dust from its internal walls and take it away (removing fungal mycelium with it), and then cover them with a thin layer of propolis. They have not lost this habit of processing the walls with propolis, even when settled by man in artificial "apartments" (i.e., hives). This fact alone indicates the important role of propolis in the life of bees.

Propolis is distinguished from other medicines - especially those invented by humans - by two features. Firstly, it has a very wide spectrum of antimicrobial activity. Secondly, it contains many hundreds (!) of various substances with different biological mechanisms of antibacterial action. And – I emphasize again! – this absolutely excludes the possibility of microbes developing resistance to the whole spectrum of substances acting in different ways. In this regard, the relatively small number of scientific studies and publications concerning the effect of propolis on bee pathogens is surprising. Moreover, even a few studies of this kind give very interesting results and this is a topic for reflection.

American researchers (Simone et al., 2009) treated the inner surfaces of the hives with propolis extract and in this way reduced the total bacterial "load" in it. After a certain period of time, the researchers determined the expression level of the genes responsible for immunity in young (7th day of life) bees. It turned out that it was significantly lower than that of bees from control families in which such treatment was not performed. Similar results were obtained when observing one of the species of tree ants that bring spruce gum to their nests. In the case of artificial enrichment of the nest with resin, and, therefore, suppression of pathogenic bacteria and fungi, the level of individual immunity decreased significantly (this was established by determining the antimicrobial activity of hemolymph of ants from experimental and control families).

These observations are extremely important because they indicate that honey bees (like some types of ants) have chosen a defence strategy different from non-social insects. The advantage of such a defence strategy is that it eliminates the need to "invest" in the internal immunity of each member of a family of tens of thousands of individuals. The total costs would be huge, and their price would be a decrease in the working capacity and even a decrease in the life expectancy of bees (Evans & Pettis, 2005). An alternative is the "external" defence mechanism based on the propolis of the nest, which is provided, importantly, by the efforts of a very small number of worker bees. It is known that the collection of plant balsams, in contrast to the collection of nectar and pollen, is engaged in by only a few dozen workers.

The problem of the "social immunity" of bees is still very far from being clarified. If propolis really plays a key role in it, then the question is: how? Indeed, it is believed that bees do not eat it, that is, they do not self-medicate (Simone-Finstrom & Spivak, 2010) like many animals.[28] Does it act on the immune system directly, somehow inhibiting the expression of genes responsible for immunity? Or does it act indirectly, by suppressing pathogens, the appearance of which causes the "inclusion" of these genes? How do different types of propolis affect microbes that are pathogenic for bees, or what kind of reaction do the bees themselves cause? Do these actions occur by contact, or through the gas phase, that is, with the participation of volatile emissions of propolis? These are questions that probably should be considered by experts in the field of bee biology and veterinary medicine.

Another important aspect of the relationship between propolis and the health of bee colonies concerns its effect on various parasites, such as mites, and, primarily, on *V. destructor*. Several fundamentally important questions arise: Does propolis have activity in relation to this parasite? If so, which "types" of propolis ("poplar", "birch-aspen", "poplar-birch-aspen", "red" or "green" Brazilian) are more active on the tick? In what form, and in what way can propolis or drugs based on it be used to combat varroatosis? Unfortunately, there are currently no answers to all these questions. However, work in this direction has begun and their first results are encouraging.

---

[28] *It is impossible, however, to exclude the possibility of one way or another getting a certain amount of propolis or its plant precursors into the bees. Propolis was found to induce the synthesis of an enzyme containing cytochrome P450. And this is only possible due to the internal action of propolis. This enzyme is responsible for the clearance of aflatoxins produced by Aspergillus fungi populated by pollen (Gonzáles et al. 2005; Beltrán-Ramirez et al. 2008; Niu et al. 2011).*

German scientists from the Free University in Berlin found in laboratory experiments that even short-term contact (only 5 seconds!) with a 10% solution of propolis in 55% ethyl alcohol causes the death of 100% of *V. destructor* mites. In the same contact time with a 7.5 and 5% propolis solution, 37 and 30% of the pests, respectively, died; however, mortality increased to 80 and 70%, respectively, with a contact lasting 90 seconds. At lower concentrations (1–2% propolis solution in 55% ethanol), parasite mortality decreased to approximately 15% and depended little on the duration of contact. However, even such weak solutions have a noticeable effect on mites, causing metabolic and narcotic effects (Garedew et al., 2002).

Similar results were later obtained by Argentinean researchers (Damiani et al., 2010). They found that spraying with a 10% propolis extract does not have a harmful effect on bees, but it kills 78% of the parasites. At the same time, it was shown that feeding sugar syrup with propolis supplementation does not have a positive therapeutic effect in the case of varroatosis. Moreover, high propolis syrup was toxic to bees.

Thus, the question posed above can be answered in the affirmative: propolis has a killer effect on the *V. destructor*. In addition, it turns out that the effect of propolis increases with increasing temperature. Placing the mites for only 30 seconds on filter paper, impregnated with a relatively weak, 4% propolis solution, killed 100% of the pests at a temperature of 40 °C (Garedew et al., 2003). This result is in good agreement with the observation of the scientist and the beekeeper S. Popravko: "*... when the hive is in the sun, the Varroa jacobsoni*[29] *mite does not feel particularly well in the bee family. This also applies to other pathogens that cause other diseases, such as European Foulbrood or nosemosis*" (Poravko, 1982).

The same group of German researchers tested the effect of propolis on a wax moth (*Galleria mellonella*), another pest that sometimes causes great trouble. As known, its larvae feed on wax and destroy the honeycomb. It turns out that immersion for 30 seconds in a 4% propolis solution leads to the death of all young larvae of this insect. Older larvae covered with a denser cuticle died at a concentration of propolis of 8–10% (Garedew et al., 2004).

There is currently no answer to the second of the questions posed above since German scientists used only two propolis samples in their studies. One of them was from the experimental apiary of the Institute of Zoology of the Free University of

---

[29] *It was later revealed by Anderson and Truman (2000) that Varroa jacobsoni is not a parasite of a honey bee, and that another, more dangerous species belonging to the same genus, V. destructor, is rampant in Europe.*

Berlin, and the other was from Ethiopia, the birthplace of one of these scientists. However, it can be assumed that propolis of different "types" and differing in chemical composition, or their mixture, will act on the mite in different ways.

In a recent publication (Pusceddu et al., 2019), a relationship was reported between the total polyphenol content of propolis stored by families and the level of their *Varroa* damage. Propolis from hives with a high degree of parasite infestation contained fewer polyphenols than in the case of families in which there were no parasites. The authors of this paper also noted that the stress caused by *Varroa* led to an increase in the number of female workers in the family who deliver herbal balms. The use of this material can be considered as a kind of self-medication of bees that is not associated with the intake of medication inside, *per os*.

The above data indicate that propolis plays a very important role in the life of bees. In light of this, it is worth considering the following question: Are beekeepers and breeders themselves not guilty of the misfortunes that now have to be fought against with the increase in the incidence of bees and the phenomenon of their mass death? It seems to me that two reasons played an important role. First, this is a successful trial of breeding a line of bees collecting as little propolis as possible, which "... is the beekeeper's curse" (Hoyt, 1965), since it complicates the work, negatively affecting economic indicators. Another reason lies in the growing demand for propolis, which began after the discovery and wide promotion of its medicinal properties. That is, at first man, to a certain extent, "weaned" the bees off collecting vegetable balms, and then he began to rob beehives that were already poor in propolis in order to market this desired product. As a "compensation" for the shortage of this natural antibiotic, which has unique and irreplaceable properties, beekeepers began to widely use (they were forced to do this!) chemical protection.

From this it follows that the beekeeper should in no case deprive the bees of propolis, this important remedy for the survival of the family. On the contrary, he/she must take care that his/her bee families always (especially in the spring) have a necessary supply of propolis. Under no circumstances would I urge the complete abandonment of collecting propolis for the needs of humanity suffering from all sorts of diseases. But it is worth considering a rational solution to the problem, which undoubtedly can be found. For example, for these purposes, it is possible to specifically distinguish several families and create suitable conditions for them – the accessibility of woody plants rich in balsamic secretions, which were discussed in the previous section. It is good if the bees collect different types of propolis, not only the common "poplar", but also the "birch" and "aspen" types. And, of course, it is better to use bees of the Caucasian race, which are distinguished by a special zeal for procuring propolis. On the contrary, some kind of help should be provided to bees belonging to races that are poorly manifested in this sense. Such help may consist of periodic artificial treatment of the internal surfaces in the hive with propolis extract.

We are hoping for specialists in the field of veterinary medicine to speed up the development of drugs based on this natural antibiotic and how to use them in beekeeping. So far, we have only known one such product manufactured in Austria called Beevital Hive Clean (Howis & Nowakowski, 2009), also known as BienenWohl®. According to the description, it contains: sucrose (29%), propolis (2%), oxalic acid (3.5%), citric acid (1.6%) and essential oils (0.05%). This drug does not cure bees of varroatosis, but stimulates their desire for self-cleaning, during which the mite falls off. According to the results of the studies cited above by the authors in Poland in 10 families, the efficiency of parasite removal during triple treatment with the drug was at an average level of 91.6%, with fluctuations for individual families ranging from 85.3 to 100%. It is accepted that an acceptable level of effectiveness of a particular drug should be at least 95%.

As long as there are not too many such drugs, and they are not too effective, you can use the advice of experienced beekeepers who use "home remedies" to combat varroatosis. S. Popravko (1982) used the original method for this purpose. It consisted of burning the waste from the collection of propolis and the bees were fumigated with this smoke. According to the author, this gave good results.

9.3.4. Essential oils and other natural remedies

It has long been known that plant tissues contain many compounds with strong bactericidal properties, outside of only relatively high molecular weight compounds (such as flavonoids and other polyphenols), which were discussed in the previous chapter. As a rule, lower molecular weight compounds are volatile, less polar, and can be isolated from different parts of plants by steam distillation. Plant essential oils are obtained mainly in this way. Their main consumers are the food and pharmaceutical industries. In the latter, they are used in the greatest quantities to obtain cosmetic products, but they are also included in many dosage forms. And, last but not least, because of the biocidal compounds contained in the oils, including those active against harmful fungi such as *Ascosphaera apis* (Boudegga et al., 2010).

Essential oils also contain natural insecticides and repellents. Therefore, it is not surprising that almost desperate beekeepers are increasingly turning to natural remedies to fight the diseases of their wards and benefactors, and many have succeeded in this. Some beekeepers use dried herbs (wormwood, elderberry, wild rosemary, etc.), laying them out under a ceiling canvas, and others spray with decoctions and herbal infusions or fumigate families with fragrant smoke.

An original way of using natural remedies for varroatosis is "dusting" the bees with finely ground powder from dried pine needles. It can be assumed that, in this case, a positive effect is achieved not so much due to the medicinal properties of the "preparation", but because the "powdered" bees are forced to vigorously comb it out of the body. At the same time, some part of the ticks is removed. At least the same "curative" effect has ... powdered sugar (Aliano & Ellis, 2005). The authors, using the "Be go" repellent, drove all adult bees into a special chamber attached to the tap hole, and then treated them with 225 g of powdered sugar. Of course, after such treatment, you will inevitably scratch yourself! In an experiment with 28 families, the average tick removal efficiency was 76.7 ± 3.6%.

This sweet "preparation" is safe for bees, and the result of its use is not bad, although not ideal. Perhaps another factor is at work here — the adhesion of a fine fraction of the powder to the suction cups of the legs of the mite, which causes the parasite to lose its ability to fix on the bee and fall off. Therefore, it is not so important what to sprinkle on the bees, just not to poison them. Starch, flour, and pollen ground into dust are suitable for this purpose.

Several years ago, in one of the apiaries in Colorado, in the western United States, I had the opportunity to observe a «master class" for local amateur beekeepers of this kind of Varroa disposal, and since that time I have been haunted by the question: how could its efficiency be improved? Perhaps this could be achieved by adding to the sugar small amounts of a substance that causes anxiety to the mite? The attentive reader will surely remember that octanoic acid, one of the main volatile substances of royal jelly, has this effect.
In section 3.4, I suggested that the Varroa female is not eager to enter a queen's cells containing large quantities of royal jelly, because it is intimidated by its smell. It enters the cells with drone brood only after the smell of royal jelly (and octanoic acid) has long disappeared. Maybe some of the Readers will take the trouble to test this idea? Octanoic acid is cheap, and for 250–300 g of powdered sugar, it seems to me, just a few drops of it will be enough.

Let's return, however, to herbal remedies. Scientists also did not stand aside. In recent years, many studies have been carried out on the effect of essential oils of plants on pathogenic organisms of bees. From all the varieties of these products, the first to be selected for study are those in which chemists have previously identified compounds with appropriate properties: antibiotic or repellent. In some works, the action on parasites of essential oils of many plant species belonging to different families is described. For example, in the work of Iranian authors (Ariana et al., 2002), the study of the varroacid properties covered oils of 17 species of aromatic plants. Acetone solutions with definite concentrations (0.05–2%) of essential oil in an amount of 1 ml were applied to filter paper, and after complete evaporation of the solvent, 40 *Varroa* mites were placed on it. Three hours after the start of the experiment, the authors determined the mortality of the parasites. It turned out that the oils of seven plants acted on them most of all, and all of them, with the exception of garden dill (*Anethum graveolens*), belong to the *Lamiaceae* family, which was previously called the *Labiatae* family.

The oils of lavender (*Lavandula officinalis*), oregano (*Origanum vulgare*), rosemary (*Rosemarinus officinalis*), savory (*Saturea hortensis*), green mint (*Mentha spicata*) and *Zataria multiflora* growing in western Asia and southern Europe were distinguished by a strong acaricidal effect. Savory stood out among them, the oil of which caused

90% death of mites, even at a solution concentration of 0.05%. That is, only about 0.5 mg of oil was applied to the filter paper! A solution with a concentration of 0.1% killed all parasites. So, what are all these representatives of the luciferous characterized by? A very high content of two isomeric phenolic compounds: thymol and carvacrol.

These natural compounds are remarkable for several reasons. First of all, being biocides, they are nevertheless not too toxic to humans (which is why many people like and willingly use spices from plants of the *Lamiáceae* family). It also turns out that each of them, or both together, help antibiotics to cope with microbes that have managed to develop resistance to them. This is another example of synergy.

Of course, not only the essential oils from the mentioned plants exhibit such properties. Natural products that can repel parasites are also of great interest. In testing 32 samples, a repellent effect against *Varroa* was discovered in 22 types of essential oil (Kraus et al., 1994). The most active of these was citronella oil, which is obtained from the grass plant cymbopogon (lemon sorghum). This oil is classified as a biopesticide by the US Environmental Protection Agency (EPA) and is classified as harmless to humans by the US Food and Drug Administration (FDA). In the United States, it has long been widely used to repel mosquitoes.

Among the 32 samples studied by Kraus et al., two samples of essential oils turned out to be attractants for *Varroa*. These are cinnamon and clove oils. Later studies showed that clove oil not only attracts but also kills mites with 96% effectiveness (Gashout & Guzmán-Novoa, 2009). Its biocidal action is likely due to the high content of another phenolic compound, eugenol (Fig. 9.2).

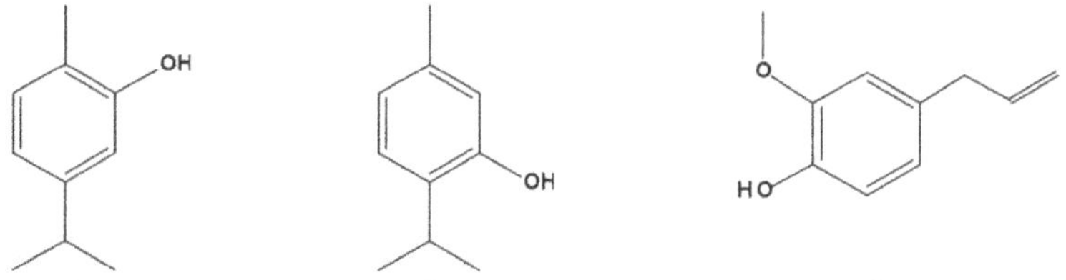

1 2 3

Figure 9.2. Chemical formulas of the main phenols of essential oils of plants: thymol (1), carvacrol (2) and eugenol (3).

The discovery of the varroacid properties of thymol and its related phenolic compounds led to the production of drugs to fight the disease. Polish beekeepers are quite familiar with the preparation Tymowarol (manufactured by VET–AGRO), produced in the form of a gel, the advantages of which were described in the Polish magazine "Apiary" at the beginning of 2007. These include the fact that "... thymol does not have a toxic effect on bees ". Is it really? Both the research results and the experience of beekeepers do not quite agree with this statement. The detrimental effects of high concentrations of thymol on bees has been pointed out for a long time (Gal et al., 1992). It was later found that, out of 22 natural products tested, thymol had the lowest half-lethal dose (LD50) for both adult bees and larvae: 210.3 µg/bee and 150.7 µg/larva. In comparison, the LD50 of menthol was more than double (382.8 µg/larva) (Gashout & Guzmán-Novoa, 2009). In addition, thymol intoxicates not only *Varroa* but also the bees themselves! Many beekeepers were convinced that after treatment with ApiLive Var, bees become lethargic, reluctant to take food, and do not guard the entrance. Worse, in some cases, this treatment leads to flight (often panic!) of bees or to the fact that they die in the hive from suffocation.

However, these tragic incidents are largely due to the misuse of the drug. If the volume of air in the hive during the use of ApiLive Var (or another preparation based on thymol) is large enough, if the conditions for good ventilation are provided, and the column of mercury does not creep up to 30 °C, then it does not lead to either the escape of the bees, or, even more, to their death.

True, it is still impossible to avoid the temporary "daze and numbness" of the bees, but it will soon pass. Moreover, using Apiguard® as an example, it has been found that increasing the free space in the hive increases the effectiveness of the drug (Lodesani & Costa, 2008). So, be careful dear beekeepers when using new medicines and strictly follow the instructions for use or seek advice from veterinarians!

Further research may lead to the discovery of essential oils with high acaricidal potential and, at the same time, that do not exhibit the undesirable side effects characteristic of preparations containing thymol. For example, some monoterpenoids, which are much less toxic to bees, have a pronounced effect on *Varroa* (Fassbinder et al., 2002).

In conclusion, a few words about the possibility of treating bees by the *per os* method, that is, by feeding sugar syrup with the addition of natural medicines, namely a few words, since there are few reports on the use and effectiveness of this approach. The already mentioned company Vita Europe Ltd. offers two preparations for medical feeding of bees: Vita Feed Green and Vita Feed Gold. The first of them is a liquid supplement containing essential oils of walnut, thyme and marjoram. The last two, as we remember, are rich in thymol and carvacrol. These two isomeric phenolic compounds are poorly soluble in water, but readily soluble in fats. Therefore, it can be assumed that the therapeutic effect of such feeding is due to the accumulation of lipophilic phenols in the fatty body of the bee, which, as we recently learned, feed on the *Varroa* mites.

Vita Feed Green is also recommended to protect bees from bacteria and fungi (including the causative agent of stone brood). Vita Feed Gold contains beetroot extract and molasses, a by-product of edible sugar production. This dressing is recommended by the company to combat nosematosis. I confess that I am not clear on what the therapeutic effect of Vita Feed Gold is based on, and I could not find documented confirmation of the effectiveness of both drugs with proper control and statistics. It is clear that there is no such data either on the effectiveness of various "home" remedies that have been widely used in recent decades by beekeepers in different countries at their own peril and risk. However, many of them reported marked improvement in the health of their families after feeding them sugar syrup with the addition of essential oils or extracts from medicinal herbs. This effect, as we recall, was observed back in the 1930s by N.P. Ioirish (see section 7.2).

Apparently, the therapeutic effect of essential oils is achieved, as in the case of thymol and carvacrol, through the transfer of part of the lipophilic (and toxic for *Varroa*) terpenoids into the fatty body of the bee. Indeed, after feeding bees sugar syrup with the addition of oregano oil or liquid protein supplementation with the addition of a mixture of oils from oregano, cinnamon oil and thymol, the larvae raised by them began to release thymol and carvacrol into the gas phase (Sammataro et al., 2009), and this can scare female *Varroa*.

In beekeeping practice, the essential oils of plants are used mainly in the autumn feeding of bees. Pure oil is added to sugar syrup (at the rate of 1 ml/l) or to candy (1 ml/kg). Here is an incomplete list of widespread medicinal plants, the essential oils of which are used by many beekeepers: peppermint (*Mentha × piperita* L.), medicinal comfrey (*Symphytum officinale* L.), horse sorrel (*Rumex* L.), bitter wormwood (*Artemisia absinthium*), field horsetail (*Equisetum arvense*), and St. John's wort (*Hypericum perforatum*).

9 3.5. Biological methods of dealing with varroatosis

Ecology, one of the subjects of which is the study of patterns in the interactions of communities of living organisms with each other, teaches us that for each biological species at least one, and more often several other species, capable of limiting its population can be named. These are their natural enemies. There are no exceptions to this rule that would lead to unlimited expansion of any species in nature, and there cannot be (truth be told, the exception, perhaps only an apparent one, is the species *Homo sapiens*, but the discussion of the consequences of its violation of the laws of nature comes out outside the scope of this book).

What about the *Varroa* mite? Based on the laws of ecology, it should be concluded *a priori* that it also has natural enemies in the places of its initial distribution. The sadness is that, setting off on its, alas, victorious campaign across countries and continents, *Varroa* did not take them with him. This is a common situation seen with almost all invasive species.

So, what kind of enemies are these – some kind of pathogenic microorganism (bacterium, fungus, or maybe virus)? Or some kind of predator? I don't know anything about the diseases that (probably) *Varroa* suffers from, but as for predatory enemies, there are candidates for this role. These are the so-called pseudoscorpions (also known as a false scorpions), small arthropod arachnid insects of the order *Pseudoscorpionida*, which got their name from their long pincer-shaped tentacles that make them look like real scorpions. One of the fathers of experimental physics, the English scientist and designer Robert Hooke, called them "land crabs".

These predatory animals are widespread in nature, but are hardly noticeable due to their secretive lifestyle and small size, rarely exceeding 3–4.5 mm. They feed on ticks, bugs and various other small insects, but they can attack organisms that are larger than themselves. Seizing the prey with their pincers, at the ends of which are exits from the venom-containing chambers, they paralyze it, and then inject digestive enzymes into the victim's body (just like *Varroa* mites when feeding on the fatty body of a larva or an adult bee).

Pseudoscorpions are found in forests under the bark of trees, in deciduous litter, and under rocks, as well as in various buildings. In our homes, the "book scorpion" (*Chelifer cancroides*) is a fairly common resident that can be found on the shelves between books, but (for the most part) lives in libraries, where it feeds on paper-eating insects. In total, there are about 3400 species of pseudoscorpions belonging to 400 genera, seven of which are called mellitophilous, that is, they like to settle with bees. These are the following genera: *Chelifera* (the "book scorpion" belongs to this genus), *Ellingsenius* (*E. ugandanus*, *E. indicus*, *E. sculpuratus*, *E. hendrickxi*, *E. pertusulatus*, *E. globossus* and *E. fulleri*), *Heterochernes* (*H. novozealandiae*), *Neobisium* (*N. validum* and *N. muscorum*), *Thalassochernes* (*T. taierensis*), *Nesochernes* (*N. gracilis*) and *Paratemnoides* (*P. pallidus*). As follows from the second part of the binomial names of some of the 14 named species, they were discovered in different countries, including Uganda, India, and New Zealand (Gonzales et al., 2008; Fombong et al., 2016), but this does not mean that they are found there and only there. For example, the "book scorpion" is a real cosmopolitan. The species I listed were found in honey bee hives, both eastern (*A. cerana*) and western (*A. mellifera*). Another six species of pseudoscorpions belonging to the genus *Chernetidae* (*Chernes cimicoides*, *Corosoma sellowi*, *Dasychernes inquilinus*, *D. panamensis*, *D. roubiki* and *D. trigonae*) were also found in the nests of stingless bees, melipone.

In world literature, the first mention of the connection of pseudoscorpions with bees in Europe appeared as early as 1873 (cited by Kistner, 1982), and in the articles of the German scientist Alois Alfonsus (Alfonsus, 1882; 1922), they were called enemies of ticks and braula (*Braula coeca*, also called bee lice). Since then, however, pseudoscorpions have rarely come to the attention of bee scientists, although some reports may have stimulated research into their role in the bee colony. For example, Indian scientists reported as early as 1947 that they were present in *A. cerana* colonies and that these colonies were free of mites and wax moths (Singh and Venkataraman, 1947).

This information was confirmed later, and the species belonging to the animals found in the families of the eastern bee was established. They turned out to be the pseudoscorpions *Elligsenius indicus* (Subbian et al., 1957). Moreover, it was found that, in the life of these animals, as well as in the life of *Varroa*, there is a phoretic phase. In other words, they "travel" with the bees, clinging to their legs with ticks (Murphy & Sudarsaman, 1986; Sudarsaman & Murphy, 1990). Obviously, in the works of these scientists, a photograph of a pseudoscorpion eating the *Varroa* mite appeared for the first time. It was called by them the "insatiable devourer" of this parasite. In Nepal, local beekeepers refer to them as "beekeeper friends" (Berube, 1999). It is not particularly surprising that pseudoscorpions were also found in colonies of *A. mellifera* imported western bees (Donovan & Paul, 2015).

The Colony Collapse Disorder (CCD) syndrome and its clear connection with the spread of varroatosis made some (unfortunately few) researchers remember pseudoscorpions as a natural enemy of *Varroa*[30]. In recent decades, there has been reliable evidence of their hunting for this parasite. Those who wish can be convinced

[30] *In a review article devoted to the prospects for biological control of Varroa, pseudoscorpions are not even mentioned (Chandler et al., 2001).*

of this by watching the film by Torben Schiffer, which shows how the "book scorpion" *C. cancroides* copes with three mites (http://youtu.be/y1zdancXRDg).

This was also reported by New Zealand scientists (Fagan et al., 2012; Read et al., 2014; van Toor et al., 2015) who studied the behaviour of *C. cancroides* and local pseudoscorpions, *Nesochernes gracilis* and *Heterochernes novazealandiae*, found in commercial hives. In the course of a laboratory experiment, the hunting of the first two species on *Varroa* was recorded using video surveillance. Moreover, a field experiment was conducted in which six adult females and one male *C. cancroides* were placed in a special container for 53 days in a hive. In the DNA of the predators subjected to this experiment, molecular biology methods (i.e. polymerase chain reaction, PCR) revealed DNA fragments of *Varroa* (van Toor et al., 2015), which were not in the DNA of the control specimens. Thus, the authors have demonstrated the possibility of reliable confirmation of the destruction of this parasite by pseudoscorpions in natural conditions without the need for visual observation (very difficult) of hunting for it.

However, some recent studies on the effectiveness of biological control of this parasite using pseudoscorpions have yielded opposite results. For example, laboratory observations of the behaviour of the mellitophilic pseudoscorpions *E. indicus* collected from *A. cerana* colonies in the Himalayas (Nepal, near Kathmandu) showed that they did not hunt for *Varroa* and were not interested in wax moth larvae, but fed on dead bees (mainly crushed hive covers), their dead larvae and psocids (Thapa et al., 2015). The authors' conclusion that *A. cerana*-associated pseudoscorpions are unsuitable for the biological control of varroatosis contradicts earlier observations (Murphy & Sudarsaman, 1986; Sudarsaman & Murphy, 1990) and seems too categorical. It should be also noted that the specific name of the predator, *E. indicus*, appears only in the title of the cited article, while in its text, no confirmation of its belonging to this (or any other) biological species is given. It is quite possible that representatives of some of the pseudoscorpion species found in bee colonies do not show a desire to hunt *Varroa*, but this does not mean that the search for the possibility of biological control using at least the well-proven *E. indicus* species is futile.

An important question arises: is it possible to breed these potential beekeeper assistants? Also, one more equally important question arises: how many individuals of this predator should be in a family to prevent an exponential increase in the number of the parasite? New Zealand researchers tried to answer both these questions. Using the example of two species of mellitophilic pseudoscorpions (*N. gracilis* and *H. novaezealandiae*) inhabiting their country, they showed the possibility of breeding these predators, on which aphids and fruit fly larvae want to feed. The optimal temperature for breeding offspring was 18 22 °C (Read et al. 2014).

Under laboratory conditions, individual helippers collected in local hives destroyed from one to nine ticks per day (they were active day and night) (http://www.youtube.com/wath?v=-qw3eVjQPXQ). Video surveillance showed that they attacked *Varroa* on brood frames, but ignored the bee larvae (Fagan et al., 2012). To estimate the number of predators required to control the parasite population in the hive at the stage of its exponential growth, the authors used a simple mathematical model:

$$N_{i+1} = N_i + \Delta - (N_c \times N_p) + C,$$

where *Ni* is the number of parasites on the *i*-th day; Δ is the change in the population (i.e. $N_i \times P$, where *P* is the rate of increase in the number of parasites per day); N_c is the number of predators (it is assumed that it remains unchanged); N_p is the number of parasites caught by the predator per day; $i = 1, \dots n$ days, and *C* is the number of parasites additionally brought by bees from other hives (it is believed that this "drift" is 1–3 individuals per day).

Assuming that there are 1000 parasites in a family of 10,000 bees and the rate of their destruction by each of the pseudoscorpions is constant and is only two individuals per day, the model predicts the number of predators required to contain the growth of parasites equal to 25. Not too large, not true is it? If the appetite of pseudoscorpions increases and they catch not two but three or four *Varroa*, the mite population will decline. However, at present, if pseudoscorpions are found in the families of *A. mellifera*, it is usually no more than two specimens. For this reason, the Ukrainian scientist I.A. Akimov in his book (Akimov, 1993) wrote that pseudoscorpions "... can capture mobile female *Varroa* mites. However, they do not restrain the growth of the parasite's population due to its high development and reproduction rate and the low number of pseudoscorpions in the hive".

The question is, why are there so few pseudoscorpions in the hive? The size of the population of one or another organism depends on the availability of food resources, as well as on environmental conditions, to a greater or lesser extent, favourable to the "everyday requirements" of this species. First of all, it is related to the breeding of offspring. Since there is no shortage of food resources in our hives (even despite the widespread use of acaricides, the number of parasites remains large), it remains to look for the reason in the poor living conditions of the predator.

The most reasonable explanation for this negative phenomenon was given by Barry Donovan and Flora Paul (Donovan & Paul, 2015). Pseudoscorpions have always looked for secluded spots for their web-woven nests, in which they lay their eggs, breed and spend the winter. In traditional hives, woven from reeds or straw, or in wooden logs, there were plenty of all kinds of irregularities and cracks suitable for these purposes. However, about 130 years ago, such primitive hives completely disappeared and were replaced by structures made of planed boards, and there were no longer places in which pseudoscorpions could build their nests. It is possible that the unintentional "expulsion" of these small predators was one of the main reasons for the outbreak of the infestation of tracheal mites *Acarapis woodi*, which led to the death of thousands of bee colonies in Great Britain at the beginning of the twentieth century (Alfonsus, 1922). This explanation of the loss of useful companions of our bees is supported by reports from South Africa, according to which there were sometimes many pseudoscorpions in recently caught "wild" swarms, but they disappeared about a week after the colonies settled in modern hives (May, 1969, cited from: Donovan & Paul, 2005).

Therefore, it is not enough to learn how to breed pseudoscorpions, but you also need to take care of their comfort. This does not require a fundamental change in the design of modern hives. Probably, it would be quite sufficient to make several cuts on the inner walls of the wooden hives or insert a special narrow frame in which cut reeds, straw or shavings are placed between the mesh walls with holes of suitable sizes.

In summary, we can safely say that there is solid hope for the biological control of varroatosis. If this can be achieved, there will be no need to use "chemical methods" of control, especially since they are becoming less and less reliable due to the development of resistance in the parasite. However, the implementation of this requires a number of activities. This is, first of all, the involvement of specialists in the field of araneology for examining bee families in different regions and identifying pseudoscorpions inhabiting them. Among the detected species, it is necessary to select the most likely candidates to control varroatosis and exclude potential harm to bees from their side. It cannot be ruled out that the best "controllers" will be not the local species of these predators or the cosmopolitan "book scorpion", but those that are found in the families of *A. cerana* and who have hunted *Varroa* for many centuries, if not millennia. That is, we are talking about attracting his primordial enemies to fight the parasite, whom he forgot to take with him on his trip around the world. On the basis of the identified "hunting" characteristics of individual species, it is necessary to determine the number of predators required to combat parasites. Of course, it is also necessary to develop a technology for the reproduction and transport of pseudo scorpions, which should be as commercially available as nucleus of bees or queen bees.

It seems that biological methods of controlling *Varroa* invasion can also develop in two other directions:

– the use of certain signalling compounds (pheromones, attractants, or repellents) to attract the parasites into traps and then destroy them or scare them away,

– search for microorganisms (entomopathogens) specific for *Varroa* that do not harm bees and their brood,

– breeding work aimed at selecting populations of *A. mellifera* bees with the same high level of "personal hygiene" as in *A. cerana* bees.

Actually, searches in both these directions are underway, but, unfortunately, have not yet brought success. I managed to find only two patents concerning the first of them (Arnold et al., 1992; Erickson et al., 2008), which is not much, especially against the background of many dozens of patents for various methods of using chemicals to combat varroatosis. In addition, both of these patents have not been practically implemented in the form of any drugs.

With regard to microbiological control, there are reports of a test as an agent against *Varroa* of the well-known entomopathogenic fungus *Metarhizium anisopliae* (Metschinkoff), isolated from the beetle *Anisoplia austriaca*. This microscopic fungus is widely used as a biopesticide. In nature, there are a large number of its isolates (strains) adapted to certain types of insects. When the spores of the fungus come into contact with the body of the host insect, they germinate, forming hyphae that penetrate into the cuticle. The insect is killed in just a few days, thanks to the production of insecticidal peptides by the fungus. Two isolates, Qu-M845 (Rodriguez et al., 2009) and BIPESCO 5 (Ferrari et al. 2020), have been tested as potential control agents for *Varroa*. In both cases, the bees were dusted with conidia of the fungus. After treatment with dry conidia sprinkled on frames and between frames, it turned out that bees were infected with mites by 65–67% less than in the control group. It is important that no negative effects of such treatment on bees were observed. These results demonstrated the feasibility of developing a biological insecticide as an alternative to *Varroa* control. It is possible that a higher efficiency can be achieved using isolates from other members of the *Acari* genus, to which *V. destructor* belongs, for example, isolate F506 from *Boophilus* ticks.

There is ample evidence of significant differences in honey bee colonies in their response to pathological factors (Seeley et al., 2015; Locke, 2016). All families die in the apiary, except for one or two, and these surviving bees are cheerful and active! The beekeeper would be wise if, when restoring the apiary, use the larvae from these colonies to raise queens for new colonies. It is possible that the beekeeper will be able to increase the resilience of families, although the success of such "empirical" selection, of course, cannot be guaranteed.

There is also no guarantee that "scientific" breeding will quickly bring comforting fruits, providing us with *A. mellifera* bees that are able of coping with *V. destructor* as successfully as *A. cerana* bees. However, these two species (the parasite and the host) have come a long way of co-evolution, resulting in a kind of balance being established between them. The question is, how far will breeders be able to shorten this road? It seems that the prospect of a quick selection of *Varroa*-resistant bees, which seemed quite real at the beginning of the millennium (Wilde & Koeniger, 2000), has not disappeared without a trace but moved into an unknown distance. Work in this direction continues and is fortunately not without success. For example, honey bee lines have been reported with an inherited ability to suppress *Varroa* reproduction by selectively removing, from larval cells, only those female ticks that are capable of laying eggs (Harbo & Harris, 2005; Harris, 2007).

An interesting experiment was started (and has been going on for over 20 years) by Swedish researchers. In the late 1990s, 150 bee colonies collected from different localities in Sweden and representing different races of bees were brought to the island of Gotland in the Baltic Sea. Previously treated families were artificially infected by replanting the same number of *Varroa* mites. The aim of the experiment was to track the survival rate of parasite-infected families neither subjected to any treatment nor measures to prevent swarming. In the first 3 years, more than 80% of families died, which was caused by a rapid increase in the number of parasites. However, after these initial losses, mite numbers declined in the surviving colonies and swarms resumed, which almost ceased in the third year of the experiment when the families became very weak (Locke & Fries, 2011).

The results obtained indicate the ability of the *A. mellifera* bee to independently (without the slightest human intervention!) adapt to the presence of *V. destructor* in the family. This adaptation did not consist of the intensification of hygienic behaviour (removal of pests together with the affected larvae) or in a more thorough self-cleaning of parasites, which is characteristic of *A. cerana*. The adaptation strategy was different. The surviving colonies in the summer period contained fewer adult bees; they had about half the bee brood and about ten times less drone brood than usual. In addition, the colonies were actively sharing by swarming. Naturally, all of this has a bad effect on the productivity of families. In short, the surviving bees have lost almost all the qualities that were grafted into them as a result of painstaking breeding work that lasted for decades!

The results of the Swedish experiment can console bee lovers who are concerned about the fate of these organisms. *Our bee can still cope with the misfortune brought on by humans*! However, if the survival rate of *A. mellifera* without treatment is accompanied by the loss of its "productive characteristics", then this cannot please the professionals operating within the existing paradigm of beekeeping. They do not need such bees, and they will continue to resort to the traditional treatment with acaricides. While those are acting or as long as consumers agree to buy products contaminated with acaricides.

9.3.6. Zoo technical approaches

Zoo technical methods of dealing with varroatosis are based on well-known data on the stages and dynamics of the development of this disease, as well as on the bee colony. The fact that Varroa reproduces mainly on drone brood suggests the possibility, although not of a complete cure of colonies, of at least a radical limitation of the number of parasites in them. It consists in the destruction of drone brood. On the other hand, the knowledge that in the life of a bee colony there are periods when it limits, or even completely ceases to breed, helps to find the optimal time for medical and preventive measures.

Regular and systematic destruction of drone brood can be based on the installation of an additional empty frame in the hive (between the frames with brood and feed). If this is done during the preparation of the colony for the swarm, the bees will immediately begin to build drone cells on it. The job of the beekeeper is to cut out the sealed drone brood and use it rationally (see Chapter 4).

In the United States, MiteZapper LLC offers a more "technological" version of this tick-killing technique, eliminating the need for bees to build drone-cell honeycombs. In addition, it leads to savings in honey, since bees need 8 parts by weight of honey to produce 1 part by weight of wax. The device, called the Mite Zapper®, is an electronic heater made in the shape of a frame with drone cells. It is installed in the centre of the nest among other frames and the queen lays unfertilized eggs in the cells. After 18–20 days, the device is connected to the power source for about eight minutes. It can be a car battery, or a 12 volt battery that produces 35 Ah. This generates enough heat to kill the larvae and with them *Varroa*. After about three days, the bees cleanse the cells and the queen can start laying eggs in them again. The manufacturer recommends heat treatment 4–5 times a year.

This approach gives good results, but even when using effective techniques, it is important not to overdo it and know when to stop! After all, it is not in vain that bees breed a huge number of drones, which require large quantities of valuable products to feed (i.e. royal jelly at the stage of a young larva and pollen and honey at subsequent stages).

Mating of the queen with many drones prevents the species from degenerating due to inbreeding. Therefore, the desire to produce as many drones as possible is quite natural. Prof. Kozhevnikov was one of the first to defend the drones. He stressed that limiting the natural needs of bees in a large number of drones negatively affects the state of the families. Overzealous destruction of "parasites" can be costly. There have been many known cases when, due to a shortage of drones, the queen remained infertile or began to lay unfertilized eggs due to a small supply of sperm. In addition, even complete removal of the drone brood from *A. mellifera* colonies does not guarantee 100% elimination of the mite. It is known that this parasite can also reproduce on the larvae of female workers of the honey bee, in contrast to the bees of the species *A. serana*.

To combat varroatosis, they also resort to interrupting the developmental cycle of ticks. For this purpose, in summer, conditions are created under which the family remains without brood for a certain time. As a result, the mite is deprived of the possibility of reproduction, and the bee colony is, as it were, cleansed of pests. It is possible to regulate brood numbers without stressing the colony if the queen is in the hive all the time but does not lay eggs. This is achieved by using special insulators. The Ukrainian scientist P. Khmara has developed and patented the device, which is a lattice box (slot width 4.3–4.4 mm) made of polypropylene, measuring 360 × 240 × 10 mm with round holes with a diameter of 14 mm in the centre of the side walls. The queen bee is introduced through these holes and closed with a stopper.

Under such conditions, the pheromones of the queen continue to circulate in the hive; therefore the bees (according to the author) do not lay fistulous queen cells and laying workers do not appear.

In the Ukraine, this method began to be used relatively recently, and there are still no reliable reports of its effectiveness in the fight against varroatosis. However, many beekeepers use it to stop the hatching of new bees in late summer and early autumn, as well as early spring. Judging by the reviews, such a stop leads to a decrease in feed consumption, and, which is worth paying attention to, to a decrease in the number of dead bees in spring. The latter circumstance may be due to the fact that the bees that did not participate in brood rearing leave in the winter in better physical condition. According to P. Khmara, the main "wear and tear" of the bee, its physiological aging, occurs when it produces royal jelly, and during this period the load on the body turns out to be incomparably greater than during flight work. This is contrary to conventional wisdom, but there is something in it. Indeed, a nursing bee (like a mother who is breastfeeding a baby) is forced to direct all the forces of her body to the development of a super-nutritious food.

The possibility of preventing diseases and premature aging of bees by stopping egg-laying by the Khmara method deserves attention. However, I am far from the intention of promoting it and calling for widespread and immediate implementation for different reasons. First, it is doubtful that, with prolonged isolation of the queen in the summer, the bees will not strive for its quiet replacement. After all, they continue to feed her with royal jelly and expect adequate behaviour, that is, the continuation of egg-laying. Secondly, it is not clear how the violent restriction of this natural function affects the reproductive ability of the queen bee. Nothing of the kind happens in natural conditions; the colony can reduce the number of brood under unfavourable conditions, but not in this way. Finally, the forcible cessation of egg-laying in the summer is completely unacceptable in the presence of late summer honey flow.

On the other hand, the danger of throwing the baby out with the bath water should be avoided. Under certain conditions, it may be advantageous not to interrupt, but only to restrict egg-laying after the main flow. For example, by changing the design of the insulator, which would allow placing 1–2 frames with foundation in it. Having worked out a suitable algorithm for replacing these frames, it is possible to reduce the cost of harvested stocks for feeding female workers who will certainly not participate in the collection.

9.3.7. Medication-free treatment

The first meeting of a honey bee *A. mellifera* with a *V. destructor* mite probably occurred in Russia, in the far east of the country. In the Primorsky Territory in the 1930s, large apiaries were laid, each of which had several thousand hives. On one of them, in the Spassky district of the Vladivostok region in 1939, N.P. Ioirish began his experiments on obtaining herbal honey (Chapter 7). Even then, local beekeepers had to independently find ways to combat varroatosis. This method was found, and most likely, by observing the behaviour of the mite in hot weather.

Remember, dear Reader, the advice that a mite "feels uncomfortable in a bee family" in a hive that has warmed up in the sun (Popravko, 1982). Or the eccentric who, in hot weather, acted with a sledgehammer, dislodging the mites (that were pickled from the heat) off bees. Probably, such observations prompted those beekeepers in the far east of Russia to get rid of mites by warming the bees. It happened like this: In the autumn, bees shaken out into net cages were brought into a very hot bathhouse and hung from the wire. From time to time, the cages were shaken and the beekeepers watched as the mites fell from them. Simple, cheap and effective!

And now let us place this method in a scientific basis, called heat treatment. First, how does fever affect both insects: the parasite and the host? Fortunately, it turned out differently. The *Varroa* mite begins to show great anxiety even at temperatures above 40 °C: It leaves the segments of the surface and begins to look for a place where it is cooler. At temperatures above 45 °C, it can no longer stay on the bee, and at 48 °C it falls off and dies. But what about the bees?

It is well known that bees actively regulate temperature and humidity in their nests. When a brood appears in it, they raise the temperature of the cells to 32–34 °C, but this is not the limit. The ability to thermoregulate is used by bees as a defence against biological threats. Using "heat weapons" against alien queens or wasps, they can raise the temperature in the cluster to at least 45 °C (Esch, 1960). Not such a drastic, but quite noticeable increase in the temperature of brood cells in response to the artificial introduction of *Ascosphaera apis* were interpreted by Starks et al. (2000) as an adaptation reaction aimed at preventing the spread of the disease. The bees themselves, according to Russian researchers, are able to withstand a two-hour stay at a temperature of 50 °C (some bees in this experiment survived until the 6th hour). We see that bees tolerate elevated temperatures much better than parasites spoilt by comfortable conditions.

Attempts to manufacture equipment for the anti-*Varroa* heat treatment of bees date back to the early 1970s, and the first heat chamber was made, obviously, in Japan (Kiroyazy, 1973). In the USSR, such a device appeared at the end of the same decade (Karpov & Zabelin, 1978; Khrust, 1978). Later, several more chamber options were proposed, and even their factory production was established. However, with the advent of effective chemicals that do not require much labour and time, the idea of heat treatment has receded into the background. The imperfection of the equipment also contributed to this. Nevertheless, some beekeepers in Russia have been using it for many decades, while others have worked hard to improve the design of heat chambers and the heat treatment methods themselves. In the last decade, due to the apparent deterioration in the efficiency of the use of synthetic acaricides (due to the progressive resistance of *Varroa*, as well as the development of "organic" or "ecological" beekeeping), interest in heat treatment has revived and commercially available devices for its implementation have appeared in some countries. Below I give a brief description of the most famous models.

One of them is the ЯВ 79/79 heat chamber of the Yarankin and Bogomolov system (Bogomolov & Yarankin, 2011). In the insulated case, equipped with inspection windows, are a heater, thermostats, fans and an electric drive for rotating the cassette in which the bees are placed. The temperature in the chamber is regulated with an accuracy of 0.1 °C, and the fan system provides uniform heating in its entire volume at the level of 47 °C. The device includes mesh cylindrical cassettes and a funnel for collecting bees. Their appearance is shown in a photograph provided courtesy of the inventor.

The basic rules of heat treatment are simple: It should be carried out in the autumn after the bees exit the sealed brood cell. The optimum outside temperature during the treatment period is 5–8 °C, but may be in the range of 2–10 °C. The air in the chamber should be warmed evenly, and the temperature should be 46–48 °C. The processing time of the cassette containing the bees is 12–15 minutes. The experience of many beekeepers using heat treatment is that there is no need to separate the queen. In one cassette shown in the photo, up to 1.5 kg of bees can be processed.

I will present the advantages and disadvantages of processing bees using this heat chamber. The advantages are:

– it is enough to carry out one-time (autumn) processing of families;

– a high degree (up to 100%) of the release of the family from the parasite,

– heat treatment is completely safe for bees,

– heat treatment acts on bees' wellness,

– heat treatment does not lead to contamination of bee products, which is a serious problem worldwide.

It is also important to note that, according to the reports of beekeepers practising this method, there is no need to take any additional anti-*Varroa* measures in the next season. Some of them, however, resort to the destruction of the drone brood and argue that with this approach, heat treatment can be carried out every two years. The general healing effect on bees is based on the fact that some microbes do not survive at temperatures of 40–45 °C. It is known that nosematosis causes especially large losses in areas with a cold climate and long winters. This disease is a real scourge of beekeeping in the northern regions of Russia. With great satisfaction, many beekeepers note its retreat after heat treatment. This was the reason for the following review by one of the beekeepers about heat treatment: "The benefit of it is such that even in the absence of varroatosis I would not refuse to use it!" (quoted from: Bogomolov & Yarankin, 2011).

There are no ideal methods and approaches to anything, each of them has its own limitations. In the case of heat treatment using the Yarankin chamber, the bottleneck is the relatively low productivity. As a rule, two well-coordinated people work in three families in one hour, that is, in a short autumn day, approximately 15 families can be served. Preparatory procedures take a lot of time: each frame needs to be removed from the hive and the bees shaken off into a cassette. Therefore, some Russian beekeepers have recently begun to use D. Tamakhin's device, which allows you to collect bees in cassettes directly from the hives in just 3–4 minutes. This makes it possible to pass through a heat chamber up to 30 families in one working day. The action of the device is based on the aspiration of bees from the housing directly into the cassette by the air flow, which is created by a fan located behind the cassette installed in the device housing.

Figure 9.3. Heat chamber ЯВ 79/79, a cylindrical cassette and a bee-collecting funnel.

As it turned out, not only adult bees, but also their brood relatively easily tolerate heating to critical temperatures for *Varroa* (Engels & Rosenkranz, 1992; Goras et al., 2015; Kablau et al., 2020a). It is also important that the presence of drones for three hours at temperatures that are fatal to the parasite does not affect their fertility (Kablau et al., 2020b).

This circumstance is used in a device called "Varroa Controller", created by the Austrian scientist and beekeeper Wolfgang Wimmer (Wimmer, 2019), and the method itself is called "hyperthermia". In the described device, the appearance of which is shown in Fig. 9.4, the sealed brood is processed. Currently, several device modifications are available: VC03 - Light, VC03 - Standard and VC03 - XLarge, the cost of which ranges from 2256 to 2568 euros. They are designed to handle 20 to 30 frames of different sizes.

Figure 9.4. External view of Varroa Controller, model VC03 - Light (www.varroa-controller.com).

The frames with brood removed from the hive after shaking off the bees are installed in a fully automated "Varroa Controller" and heated in a humid atmosphere at a temperature of 43.7 ° C for 2 hours (however, Professor Wimmer does not name the processing temperature either in his book or in interview)[31]. As a result, both the eggs and larvae and the adults of *Varroa* die and remain "buried" in the cells until the cured bees hatch from them. After the end of the procedure, the frames with brood are returned to the colonies.

The obvious disadvantage of this method is the need to carry out several repeated treatments of the same families during the year. Indeed, even with 100% mortality of *Varroa* in the sealed brood, there are still "phoretic" parasites in the family, which will immediately begin to restore the population, the number of which can double within a month. Moreover, the productivity of the method is very low: no more than three families can be processed in one working day.

Another device for treating hyperthermia, called the Hive Thermal System, was developed by the American company Bee Hive Thermal Industries, South Carolina. It allows you to handle the family as a whole, both adult bees and brood. The device includes a panel-shaped heater and an electronic device (Fig. 9.5).

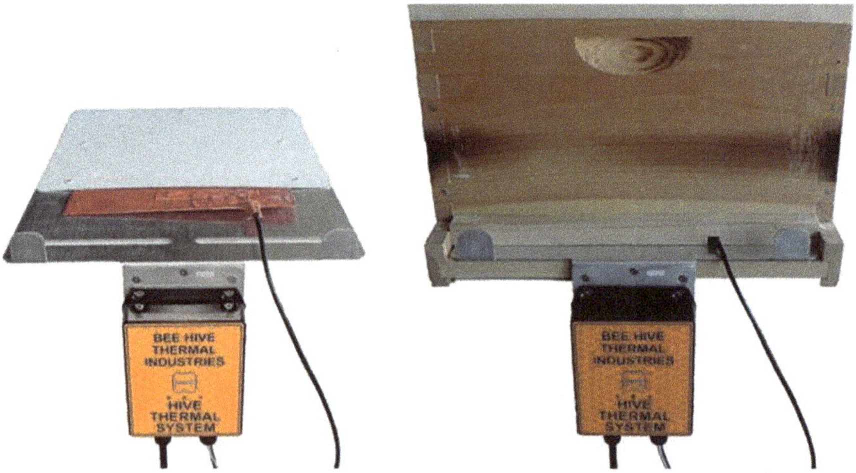

Figure 9.5. Exterior view of family heat treatment device (www.beehivethermalindustries)

The video link https://youtu.be/D3I4G2Ws91o explains how to use this commercial product. For the treatment of the whole family, brood and adult bees, a

[31] *According to Goras et al. (2015) at a temperature of 42 ° C, heating for eight hours was required to kill all 100% of the ticks in the family.*

device developed in Germany called BIENENSAUNA® is also intended. Processing is carried out by a stepwise increase in temperature to 41–42 °C, and its duration is 2.5–3 hours. In addition to the devices mentioned, others are currently commercially available, such as Borgstadter-Termo-Box, Apitherm Box, Varroa Kill II. The abundance of developed and proposed models testifies to the interest of beekeepers in the transition to the post-acaricide era.

A common disadvantage of all the mentioned models of devices for hyperthermal treatment is low productivity: no more than three families can be treated in one working day. Even the ability to treat 30 families in one day using the Yarankin thermal camera may not seem high enough. Especially for those beekeepers who have several hundred of them on their farm. But for an apiary with up to 50 hives, hyperthermia can be a kind of lifeline. And, moreover, it will be useful in farms that practice "ecological" beekeeping (in many countries such have already appeared). Usually, in such farms, the number of bee colonies is not too large, and the apiaries themselves are stationary. On the other hand, in "ecological" apiaries it is not allowed to use preparations from the arsenal of synthetic chemistry, acaricides, fungicides and antibiotics for the prevention and treatment of bees. For them, the use of thermal treatment may be the only way to combat varroatosis.

As it turned out, *Varroa*'s fear of heat is not the only weak point. The fecundity of female ticks is affected by the temperature at which the parasite reproduces. As early as 1960, it was established that short-term cooling of the brood of worker bees does not lead to a violation of their development. The situation is different with *Varroa*. In an experiment carried out in Greece (Ifantides, 2020), frames with sealed brood on the fourth day after sealing were removed from the hives and transferred to a room with an air temperature of 24–25 ° C and left in it for 24 hours. After that, they were returned to their place. Two to three days before the expected time of hatching the female workers, the brood cells were opened and examined stereoscopically in cold light for the fecundity of the "parent" females. They were considered fecund if there was at least one adult daughter in their offspring. As a result, it was found that only 3.5% of females were fertile, while in the control, their share was 85.7%. It follows from this that, in principle, not only hyperthermia, but also hypothermia can be used to curb the growth of the *Varroa* population in families!

And in conclusion, a few words about another approach to combating varroatosis. It is about using ultrasound for this in a certain frequency range. When bees were exposed to ultrasound in the frequency range from 14 to 80 kHz at an intensity level of 90–110 dB, no changes in the behaviour of bees were observed (Barry et al., 2018). However, the ultrasound greatly disturbs *Varroa* and makes it difficult for the parasite to adhere to the bee's body. This allows the bees to easily clean themselves (auto-grooming) and each other (allo-grooming) of mites that fall to the bottom of the hive. Although the method was patented over 15 years ago (Rümmelin, 2004), no reports of its practical application have appeared in the literature to date.

9.4. Bee quarantine disease: American foulbrood

9.4.1. The causative agent and the clinical picture of the disease

Bacterial diseases (bacteriosis) constantly pose a great threat to the health of bees. A disease known for at least 250 years, called American foulbrood (AFB), is of serious concern around the world. Before the appearance of varroatosis in Europe, AFB was considered the most dangerous disease of bees, causing the greatest losses in beekeeping.

The causative agent of AFB is the gram-positive aerobic bacterium *Paenibacillus larvae*, which can also grow under anaerobic conditions (the so-called facultative anaerobe). It infects only larvae of a certain age, namely on the fourth or sixth day after hatching from the egg. Adult bees do not suffer from this disease, but they serve as carriers of endospores, which are dormant forms of bacteria. The larvae become infected when they ingest food contaminated with endospores. It is believed that the development of the disease can be caused by the penetration of only 35 bacterial spores into the larva's body, and only 10 spores are sufficient to infect the youngest larvae. This means that AFB is one of the most contagious diseases known to man. Another feature of AFB is the extremely high viability of its endospores.

Under environmental conditions, they can remain active for more than 35 years, and at 100 °C, they die only after five days. The source of infection is the remains of larvae that have already died from AFB. After the death of the infected larva, its tissues disintegrate. But, even after that, bacteria continue to multiply in them, which turn into endospores after 7–10 days. Their number in the corpse of one larva can reach 2–3 billion. Working bees, which clean the combs from the remains of dead larvae, inevitably become infected with endospores and spread them in the hive, contaminating the stocks of bee bread and honey. For this reason, during epidemiological studies in specialized laboratories, honey samples are analysed for the content of *P. larvae* endospores in them, which can reach 15,000 per 1 g of honey.

Under favourable conditions, the endospores germinate, that is, they turn into a normal vegetative cell capable of proliferation (multiplication by division). The inducer of germination are sugars and amino acids in the digestive tract of the larva; germination begins in seconds after exposure to the inductor.

Book sources and scientific periodicals published so far do not give a sufficiently accurate description of the clinical picture of the disease and the pathogenic factor causing AFB. In them, for example, it is reported that this is a disease of the printed bee brood, that is, the larvae fall ill and die after the cell is sealed. In fact, the disease can proceed differently and affect the colony in different ways, depending on the genotype of the bacteria and on the behaviour of the bees.

Currently, five genotypes of *P. larvae* are known: ERIC I, ERIC II, ERIC III, ERIC IV and ERIC V. The results of epidemiological studies indicate the spread of the ERIC I genotype both in Europe and the Americas. The ERIC II genotype is found only in Europe. It is also reported that the ERIC III and ERIC IV genotypes have not been found in field studies, and they supposedly only exist in microbiological collections (Genersch, 2010). The first of them is listed in the Belgian collection of microorganisms under the number LMG 16252, and the second is listed under the number LMG 16247 (both were isolated in different years from dead bee larvae). As for the ERIC V genotype, it was isolated from a Spanish honey sample only recently (Beims, 2018) and is stored in the German collection under the number DSM 106052.

Genotypes ERIC I and ERIC II, which are widespread in Europe, differ in their pathogenic activity and the time during which they kill all infected larvae (LT_{100}). They can be divided according to this criterion into three groups: slow killer, moderate killer and fast killer. The slow-acting ERIC I bacteria only kill infected larvae on the twelfth day (LT_{100} = 12) (i.e. *after sealing of the cells* by the worker bees). In contrast, larvae infected with the ERIC II genotype had already died on the seventh day (LT_{100} = 7), that is, *before the cells were sealed.* The same happens in the case of infection with the ERIC III and ERIC IV genotypes. As a result of the action of bacteria of these genotypes, clinical signs of AFB quickly appear. There is a yellowing of the cuticle of the larvae and the appearance of a very characteristic smell, similar to the smell of rye bread (pumpernickel) common in Germany.

According to recent studies (Beims, 2018), in the case of rapidly killing ERIC III-V genotypes, the lethal time is only two days (LT_{100} = 2). The infected larva dies at the stage when it lies on the bottom of the croissant-shaped cell. Most of its body is occupied by the midgut, which is not yet connected to the hindgut. Therefore, undigested food particles remain in it throughout the entire larval stage. As a result, the cells with dead larvae do not contain faeces and mucus containing spores, which would stain the walls and bottom of the cells.

Thus, the virulence of different *P. larvae* genotypes towards larvae negatively correlates with the virulence for the family as a whole. Research in recent years suggests that the slow-killing (less virulent) ERIC I genotype is responsible for the highest number of colonies killed by AFB. Larvae that die earlier (2-7 days after infection) are easily detected and removed by worker bees as part of the already mentioned hygienic behaviour. Thus, it cannot be ruled out that the information on the absence of infection in bee colonies caused by "quick killers", that is, *P. larvae* bacteria belonging to the ERIC III and IV genotypes, as well as to the newly discovered ERIC V genotype, is erroneous. Removal of diseased and dead larvae before the development of endospores in them effectively counteracts the development of the disease in the family and thereby prevents its death. Due to the hygienic behaviour of bees, observers may not notice the first symptoms of the disease, but worker bees do not ignore the smell of larvae killed by the disease.

The threat of a sudden outbreak of AFB now hangs over every apiary. Due to the extremely strong epidemiological expansion of AFB, in many countries, there is a rule of mandatory destruction of infected bee colonies by lighting them up with sulphur dioxide or formalin and burning the dead bees along with the contents of the hive and related equipment. After burning bees, the disinfection of hives and equipment by burning them is allowed. In Poland, the approach to the problem of preventing the spread of AFB is not so radical, and besides the destruction of sick families, their treatment under the supervision of a veterinarian is allowed. In principle, only one single selective and sanitary procedure can be used for treatment—the resettlement of sick families. This method of treatment, developed back in 1769 by A.G. Schirach, consists of separating adult bees from diseased and dead brood. It includes keeping adult bees hungry (at least 24 hours of hunger strike!) in a disinfected hive on frames with foundation. The technique of double relocation can also be used with the use of a swarm-box at the first stage. All old cells and infected frames are definitely destroyed. This method gives positive results in the treatment of infected families, but does not guarantee a cure in the clinical phase of the disease.

For a long time, antibiotics (tetracycline, oxytetracycline, flavomycin, erythromycin, etc.), as well as sulphonamides, have been widely used to treat colonies infected with AFB. However, it turned out that antibiotic treatment with AFB does not give the expected effect for various reasons. First, their long-term use has led to the development of resistance to these drugs in the pathogen. As a result of such treatment, the epidemiological situation worsened significantly. According to the results of a study, the frequency of AFB occurrence in countries with legal use of antibiotics was higher than in countries where they are prohibited (Tian et al., 2012).

Secondly, antibiotics do not act on endospores, but only kill vegetative forms of bacteria (and even then not 100%). Since very few endospores are required to infect larvae, a decrease in their total number by even 90% does not prevent the development of the disease. An additional problem is the contamination of bee products with drug residues, which leads to a deterioration in the quality of products intended for human consumption.

9.4.2. Bacteria against AFB

A few words should be said about some factors of the insufficiently studied mechanism of development (pathogenesis) of *P. larvae*. First of all, the question arises: why do larvae of a strictly defined age get sick with AFB and why are adult bees not susceptible to infection by this highly invasive bacterium? After all, they are the main carriers of endospores, and it is they who transmit them to the larvae along with the contaminated food.

Some scientists believe that resistance to infection is associated with the presence of an anatomical barrier, which is an intermediate valve (proventriculus) of the stomach of an adult *A. mellifera*, supposedly trapping *P. larvae* endospores and preventing them from entering the digestive tract where they could transform into vegetative forms, causing disruption of the intestinal wall cells, then attacking the internal organs of the bee (Sturtevant & Revell, 1953). However, it is difficult to believe in the valve's ability to filter out 100% of the endospores. Some of them can overcome the anatomical barrier and enter the midgut with food. In addition, this does not explain the resistance to infection of older larvae, which still lack this barrier.

Other scientists believe that endospores do not germinate in the digestive tract of bees and are removed with faeces. This seems more likely, but I have not found any explanation why spore germination does not occur, despite the presence of inductors in the bee's stomach (i.e. glucose, the amino acid alanine, and water), which should activate this process. To understand what is the reason for such immunity, it is necessary to turn to the physiology of all participants in the process: pathogenic bacteria, larvae of different ages and adult bees.

First, let us dwell on the larvae and adult bees in the digestive tract, into which (even despite the anatomical barrier in the form of a proventriculus) endospores, that is, dormant forms of bacteria, enter. The development of the disease requires the interruption of the resting stage. This occurs 24 hours after the endospores enter the larva's body. As a result of their germination, vegetative cells of bacteria appear that are capable of metabolism and multiplication by division. The further fate of these cells depends on the characteristics of the environment.

In favourable conditions, which take place in the practically sterile digestive tract of a young larva (here it is appropriate to recall the antimicrobial properties of royal jelly, which the larvae feed on in the first 2 to 3 days after hatching from an egg), bacteria begin to multiply, feeding on sugars and proteins (i.e. honey and pollen gruel) that make up larval feed. Recent studies have shown that, at this stage the bacteria, *P. larvae* begin to secrete some metabolites with antibiotic properties. Their task is to protect the "niche" assigned to themselves by the bacteria in the digestive tract of the larva by suppressing the germination and further development of spores of other bacteria and fungi supplied with food. These are relatively simple chemical compounds, the so-called non-ribosomal peptides (NRPs), as well as low molecular weight lipopeptides called paenilarvines A, B, and C (Figure 9.6). As can be seen from the figure, these are cyclic structures consisting of several amino acids, including asparagine, tyrosine, proline, β-alanine, serine, and glycine. The structure of paenilarvines A and C differs in the structure of the R_2 substituent in the β-alanine residue. In paenilarvine B, one of the asparagine molecules is replaced by aspartic acid (R_1 = OH). These substances exhibit the properties of strong, non-selective fungicides. Another group of secondary metabolites of *P. larvae*, paenilamicines A1, A2, B1 and B2 (Fig. 9.7), have a relatively weak effect on fungi, but have bactericidal activity. As you can see, the pathogen is well-armed and does not allow competition from other microbes.

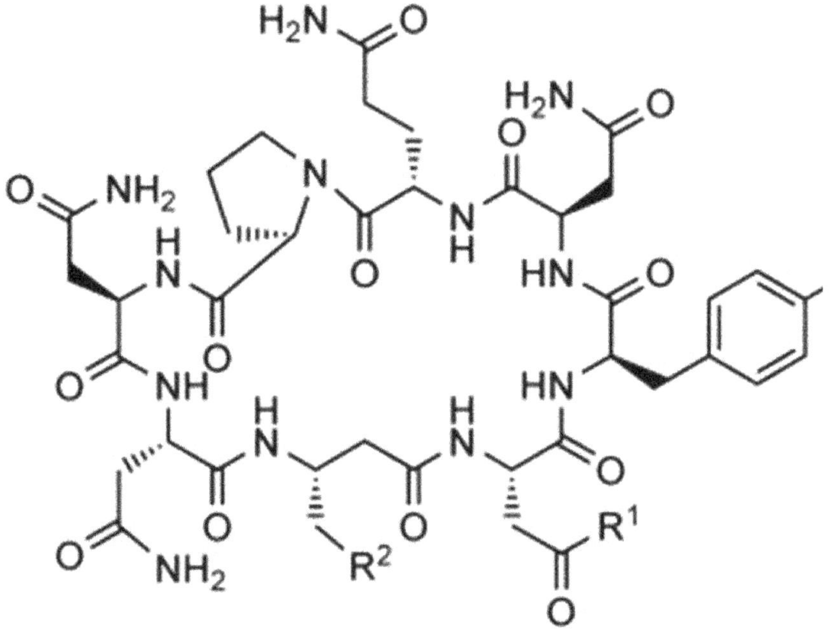

Figure 9.6. Structural formula of paenilarvines A, B and C.
paenilarvine A: $R^1 = NH_2$, $R^2 = (CH_2)_9CHCH_3CH_2CH_3$;
paenilarvine B: $R^1 = OH$, $R^2 = (CH_2)_9CHCH_3CH_2CH_3$;
paenilarvine C: $R^1 = NH_2$, $R^2 = (CH_2)_8CH(CH_3)_2$.

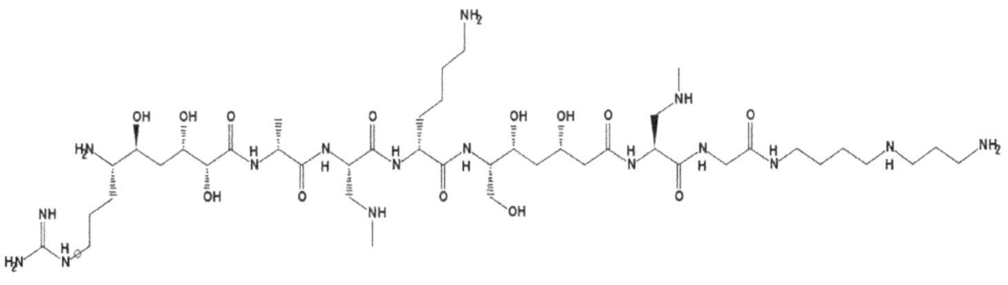

Figure 9.7. Structural formula of paenilamicine B1

At a certain stage, the growing bacteria also begins to produce the PICB49 enzyme, which decomposes the chitinous protective layer of the epithelium of the larva's intestines. Two more toxins, P1x1 and P1x2, destroy the cells of the epithelium itself, which allows vegetative forms of bacteria to penetrate into the hemocoel (the primary body cavity of the larva). At this stage, the bee larva dies and the decomposition of all its tissues begins. After the death of the larva, bacteria continue to multiply in it, which switch to the saprophytic type of nutrition, and finally decompose all tissues with the formation of a characteristic viscous mass with the smell of wood glue. The dominance of vegetative forms of *P. larvae* in it is absolutely due to the development of the above antibiotics, which destroy any "foreign" saprophytes. After 7–10 days, after the exhaustion of nutrients, the formation of endospores begins, the number of which in the remains of one larva can reach 2.5 billion.

However, the environment of the newly formed vegetative cell of the pathogen is not always as favourable as in the case of the larva at the age of 3–4 days, primarily due to acute interspecific competition. Although the bacillus *P. larvae*, as we now know, can use its own "weapon" to suppress competitors, it has no chance of success in life in the digestive tract of older larvae and, moreover, adult bees. Because, it will be met by an already established and friendly collective of bacteria living in symbiosis with the host's organism. They create the necessary conditions for their existence and protect their "niche" from its capture by competitors.

At present, it is difficult to name the number of symbiotic species of bee bacteria, especially since new, previously unknown ones are constantly being discovered. In recent years, a new technique has emerged for studying the host–symbiont relationship. To date, eight dominant groups of bacteria have been discovered, which account for approximately 90% of the entire community. The middle part of the bee's digestive tract is dominated by species belonging to the classes *Gammaproteobcteria, Betaproteobcteria* and *Alphaproteobcteria*. The posterior part of the tract (rectum) is colonized mainly by various species of bacteria of the genus *Lactobacillus* (*L. mellis, L. mellifer, L. hesingborgensis, L. kullabergensis* and *L. kimbladic*). Some of them have been isolated and identified only recently, and the study of their interaction with the bee is in its early stages.

In the course of studying this kind of relationship, scientists have long come to the conclusion that symbionts are very well adapted to the host's body and take part in its metabolism. In the case of bees, the microbes of the digestive tract perform many different functions, such as regulating and optimizing acidity; synthesizing amino acids, fatty acids and B complex vitamins; and producing enzymes needed to break down cellulose and lignin. Such enzymes secrete *Gammaproteobcteria apicola* and bacteria from the genus *Fructobacillus*, and the gram-negative bacterium *Galliamella apicola* is involved in the decomposition of the cell membranes of the pollen and the release of proteins and other components contained in them. Some of these valuable components will be used by symbiotic microbes, which protects the bees from pathogens. This kind of mutually beneficial relationship between populations is called *mutualism*.

According to many scientists, lactic acid bacteria (abbreviated as hbs-LAB), which live in the digestive tract, play a major role in protecting against many pathogenic microorganisms. To date, 43 species of such bacteria have been isolated (Ramos et al., 2019), and many of them have been tested from the point of view of action against *P. larvae*. The most active in this regard were species such as *Lactobacillus kunkei*, *L. crispatus*, and *L. acidophilus* (Kačaniová et al., 2019; Al-Ghamadi et al., 2020). Antagonistic activity against *P. larvae* is shown not only by hbs-LAB but also by some other bacteria endogenous to the honey bee, such as *Bacillus subtilis*, *B. licheniformis*, and *Fructobacillus fructosis*. The protective effect of hbs-LAB has been shown in laboratory conditions both *in vitro* and *in vivo* by feeding them to larvae. It was also found that a certain "secret" secreted by LAB acts on the pathogen (Lamei et al., 2019). The composition of this "secret" has not been investigated, but, in my opinion, the antagonistic effect is due to the main metabolite of LAB contained in it, lactic acid. This opinion of mine is based on my own experience of unsuccessful attempts to cultivate vegetative forms of *P. larvae* on slightly acidic (pH 5.1) nutrient media (Isidorov et al., 2018a).

This experience allows us to put forward a hypothesis that answers the questions at the beginning of this section. The first of them, concerning the immunity of older larvae to infection, can be answered as follows. In the middle part of the digestive tract of a young larva, immediately after it starts feeding on a mixture of honey and pollen, a microbiome is rapidly formed, consisting of symbiotic microorganisms, including hbs-LAB. Their main product of life, lactic acid, creates an acidic environment in the intestines. If endospores of *P. larvae* enter it at this time, they will not be able to germinate and turn into vegetative forms. For the same reason, endospores of a pathogen have an even lower chance of success in the stomach of an adult bee. Most likely, they die in it under the influence of the "secret" of symbionts and/or digestive enzymes of the bee.

The antagonistic relationship between lactic acid bacteria and *P. larvae* was confirmed in tests on live larvae. This gives hope for the development of new ways to combat AFB; however, the road is still not close from completely successful laboratory tests *in vivo* leading to the creation of a dosage form and method of its application. Unfortunately, the observed effects at the level of individual organisms do not always manifest themselves at the level of the whole family. For example, treatment attempts by feeding live forms of hbs-LAB to worker bees did not affect the course of the disease in the family (Stephan et al., 2019). Nevertheless, work in this direction continues, and it is hoped that they will be successful.

9.4.3. Natural remedies for American foulbrood

A variety of plant essential oils and the individual chemical compounds they contain have been tested to combat AFB. The antimicrobial effect of essential oils is attributed to the volatile and aromatic monoterpenes they contain, such as limonene, α- and β-pinenes, and others. Essential oils of different origins, such as citrus, rosemary, eucalyptus, pine and others, exhibit activity against *P. larvae*; however, their tests in *in vitro* experiments did not lead to the desired results. The minimum concentrations that inhibited the growth (minimal inhibitory concentration, MIC) of most of them are in the range of 250-400 μg/ml. Thyme and oregano oils containing thymol and carvacrol exhibit higher activity (MIC < 100 μg/ml).

Propolis is much more active against *P. larvae* (Bilikova et al., 2013; Isidorov et al., 2017). The antimicrobial effect of propolis is associated with the flavonoids and other natural phenolic compounds contained in it, which are present in various tissues of many higher plants. In this regard, the idea arose of studying the effect of extracts of selected plant objects on the AFB pathogen. As we know, the plant precursor of propolis in our latitudes is mainly resinous excretions on the surface of the buds of woody plants, such as poplar, aspen and downy birch.

The chemical composition of the resinous secretions of the buds of these plants is different, and therefore one could expect *a priori* that different types of propolis would act differently on *P. larvae*. Indeed, trials of nine propolis samples from five European countries demonstrated such differences. The development of five bacterial strains belonging to the most dangerous genotype (ERIC I) was most actively suppressed by the "poplar-type" extracts of propolis (MIC 7.8–15.6 μg/ml) from Russia and Slovakia, followed by extracts of Russian propolis "birch-type" (MIC 15.6 –31.8 μg/ml) and aspen-type (MIC 15.6–62.4 μg/ml) from Finland and Latvia (Isidorov et al., 2017).

The antimicrobial effect of propolis is associated with the flavonoids and other natural phenolic compounds contained in it, which are present in various tissues of many higher plants. In this regard, the idea arose to study the effects of the extracts of buds and other tissues of the "propolis-giving" plants of Europe, black poplar (*P. nigra*), two species of birch (*B. pendula* and *B. pubescens*) and aspen (*P. tremula*) on the AFB pathogen. The discovery of high anti-AFB activity of extracts of plant raw materials could, in principle, allow replacing them with propolis extracts, which is so necessary for maintaining the health of the bee colony.

Tests using the same strains of *P. larvae* in the case of bud extracts were very encouraging. The MIC values of the buds of "propolis-giving" plant species ranged from 7.8–31.8 µg/ml. However, a surprise was the very high activity (MIC < 1.0-3.9 µg/ml) of extracts of silver birch buds (*B. pendula*). It turned out to be about the same as the activity of tylosin, a macrolide antibiotic (Isidorov et al., 2018b), but the resinous secretions of the buds of this birch species are not collected by bees to make propolis! As we remember (and if we have forgotten, then again turn to section 5.3), the buds of *B. pendula* are very poor in flavonoids and other polyphenol (polar in nature) compounds, which are usually attributed to antimicrobial activity. On the contrary, they contain very large amounts (more than 60% of dry extract) of non-polar and weakly polar triterpenoids, in which the excretions of buds of "propolis-giving" downy birch (*B. pubescens*) are poor. Does it follow from this that low-polar (lipophilic) compounds more actively inhibit the development of *P. larvae* bacteria? Tests of the activity against them on the extracts of young branches (devoid of leaves and buds) of both birch species confirmed this conclusion. The compounds extracted with non-polar hexane acted more strongly (MIC < 1.0–3.9 µg/ml) than those extracted with weakly polar ether (MIC 7.8–31.8 µg/ml) and highly polar methanol (MIC 62.5–125.0 µg/ml).

In general, this is consistent with the observations of other authors. For example, it was noted that a decrease in the content of polar flavonoids in propolis samples was not accompanied by a decrease in their activity against *Staphylococcus aureus* (Kujumgiew et al., 1999). It was also shown that the non-polar fraction of the dichloromethane extract of the tropical woody plant *Scutia buxifolia* showed higher activity against six different *Paenibacillus* species than the extract with moderately polar ethyl acetate and polar butanol (Boligon et al., 2013). Among the 3-acylpinobankins isolated from American poplar-type propolis, homologues with longer acyl groups (i.e. less polar ones) had the highest activity against *P. larvae* (Wilson et al., 2017).

To test the relationship between the polarity of various compounds and their activity against *P. larvae*, experiments were performed with selected triterpenoids, flavonoids, and polyphenol glucosides (Isidorov et al., 2018b).

These experiments demonstrated, for the first time, the high anti-AFB activity of triterpenoids compared with more polar flavonoids, and with strongly polar glucosides. On their basis, it can be assumed that potential drugs for combating AFB should contain lipophilic fragments in the molecule, allowing them to attach to lipids of the bacterial cell membrane and/or bind to the hydrophobic part of proteins embedded in the cytoplasmic membrane and, thus, disrupt their structure.

The *in vitro* laboratory observations described above are important; however, the possibility of their application in veterinary practice can be judged only after testing on bee larvae and on adult bees. *In vivo* studies of this kind are carried out, and in some cases, their results are promising. Preliminary tests carried out at the Institute of Forest Sciences of the Technical University of Bialystok under the guidance of Prof. S. Bakier suggest that extracts of certain multi-fruit arboreal fungi may be a natural alternative to the fight against AFB. For example, an extract of one of the species of saprotrophic fungi of the genus *Sceletocutis* in *in vitro* experiments with bacteria of the ERIC I and ERIC II genotypes demonstrated MIC values of 0.3–4.8 µg/ml.

In subsequent *in vivo* experiments carried out at the Institute of Veterinary Medicine in Puławy (Poland), the larvae were fed with pure royal jelly and royal jelly containing 0.025% and 0.05% of this fungus extract. As a result, it was found that the extract in both concentrations showed low toxicity. By the end of the experiment (on the 11th day), 90% and 82% of the larvae in the control and experimental groups survived, respectively. On the other hand, by the end of the experiment, 100% of untreated larvae infected with bacteria of the ERIC I and ERIC II genotypes had died. The survival rate of larvae infected with bacteria of both genotypes and fed with food with 0.05% extract was 3% and 50%, respectively. Thus, the extract from *Tyromyces* sp. demonstrated a selective effect on the more virulent genotype of *P. larvae*.

It can be assumed that, in the case of infection of larvae with bacteria of the ERIC I genotype, in order to achieve a therapeutic effect, it is necessary to use a more concentrated extract of the studied fungus. *In vivo* experiments with extracts of this and other woody fungi will continue, and we hope to get good results.

9.5. Infestations of pests and pathogens continue

Unfortunately, the “pathosphere” of bees in Europe and North America does not stop expanding due to the ongoing “import” of pathogens from their historical habitats. An example is the appearance of a small beehive beetle (*Aethina tumida*), an inhabitant of the so-called sub-Sahara or Black Africa, first in the USA (1996), then in Canada and Australia (2002), and finally in Europe (2003). The "corridor" through which it spreads is the importing of queen bees and bee bags. However, one cannot exclude the possibility of it being imported from Africa with some other goods.

This pest is very mobile (adult insects fly up to 20 km); it is extremely tenacious in different conditions (tolerates extreme heat and drought, as well as frosts down to -40 °C). It is also distinguished by excellent fecundity: 4–5 generations of beetle are hatched per year. In laboratory conditions, a 450-fold increase in the number of parasites was recorded in just four and a half weeks! Bees cannot cope with this pest, because it is covered with a thick layer of chitin, which is impenetrable by stings, runs around the honeycombs very quickly and can hide wherever possible. If you do not fight it, the beetle migrates to a distance of up to 2.5 km per year. If you do not fight it, then the infection spreads to a distance of up to 2.5 km per year (according to this indicator, it lags behind *Varroa*, the spreading rate of which reaches 11 km in one season).

As with the *V. destructor*, this parasite does not cause much harm to the bee in its original habitat. It is not clear whether this is due to the special hygiene behaviour of African bees (*Apis mellifera scutellata* and *A. mellifera capensis*), or to the presence of some pathogen that limits the population of *A. tumida*.

It should be noted that not only beehives, but also all utility rooms where apiary products are stored, such as in frames with bee bread and frames after the extraction of honey, are also a desirable habitat for the small hive beetle. The main harm is caused by larvae that hatch from the eggs laid by the female no more than six days later. The larvae are omnivorous and extremely voracious: they feed on everything that is in the hive: wax, bee bread, honey, larvae and pupae of bees and even dead bees. After 6–10 days of reinforced feeding, larvae (at the so-called "wandering stage") get out of the hive through the entrance, fall to the ground and burrow into it to a depth of 20 cm (Neumann et al., 2016). It is noted that they prefer moist soil and, in search of a suitable place, can move up to 100 metres from "their" hive.

There are no other means to combat the invasion of *A. tumida*, except for the introduction of strict quarantine restrictions. The fact that such measures give a good result is evidenced by some examples. In Portugal, the pest was first detected in one of the beekeeping farms in 2004, but emergency measures prevented its spread. In 2014, it appeared in the southern regions of Italy, in Sicily and in Calabria. This caused great concern, as Italy is one of the main exporters of bee colonies (up to 20,000 per year).

Measures were taken to localize and eliminate the potential breeding ground for the pest, including the destruction of sick families and soil treatment with insecticides. There was also a long-term ban on the placement of beekeeping farms in the places where the pest was discovered. In general, the danger was eliminated at great cost – at least, the appearance of *A. tumida* in England, one of the main importers of bee packets from Italy, was not reported. It is hoped that this misfortune will pass other European countries by. However, it is already causing great problems in the western hemisphere. An invasion of *A. tumida* has been reported in Mexico (Reyes-Escobar et al., 2015; Valdovinos-Flores et al., 2016; Hernandez-Torres et al., 2018), which has raided beehives in the Yucatan Peninsula.

Strict quarantine measures can also save Europe and America from the spread of the new causative agent of nozemosis. In addition to two known species, *Nosema apis* and *N. ceranae* (the first of them was already known in Europe 100 years ago, and the second was "imported" from Asia, apparently only in the late 1990s), a previously unknown species, *Nosema neumanni*, was discovered in 2017. To date, it has been found on *A. mellifera* bees only in Uganda it is hoped it will not begin its journey around the world.

References

Chapter 1

Butler C. (2017). The Feminine Monarchie edited by John Owen. Yorkshire, UK: Northern Bee Books.

Crane E. (1999) The World History of Beekeeping and Honey Hunting. London, Duckworth.

Hollier J., Hollier A. (2015) Huber the Bees: Françoise Huber and the science of entomology in eighteenth-century Geneva. Antenna, 39 (2).

Tomaszewski R. (2011) Not only Dzierżon: an introduction to the bibliography of the literature beekeeping in Silesia, doi.org/10.14746/b.2011.15.2 (in Polish).

Chapter 2

Baudry E., Solignac M., Garnery L., Gries M., Cornuet J.-M., Koeniger N. (1998) Relatedness among honeybees (*Apis mellifera*) of a drone congregation area. Proc. Royal Soc. Lond. B, 265, 2009–2014.

Beggs K.T., Glendining K.A., Marechal N.M., Vergoz V., Nakamura I., Slessor K.N., Mercer A.R. (2007) Queen pheromone modulates brain dopamine function in worker honey bees. *Proc. Natl. Acad. Sci. USA*, 104, 2460–2464.

Blum M.S., Fales H.M. (1988) Eclectic chemiosociality of the honeybee: a wealth of behaviors, pheromones, and exocrine glands. *J. Chem. Ecol.* 14, 2099–2107.

Calderone N.W., Page R.E., Jr. (1988) Genotypic variability in age polyethism and task specialization in the honey bee, *Apis mellifera* (Hymenoptera: Apidae). *Behav. Ecol. Socibiol.* 22, 17–25.

Crew R.M., Velthuis H.H.W. (1980) False queens: a consequence of mandibular gland signals in worker honeybees. *Naturwissenschaften*, 67, 467–469.

Dor R., Katzav-Gozanski T., Hefetz A. (2005) Dufour's gland pheromone as a reliable fertility signal among honeybee (*Apis mellifera*) workers. *Behav. Ecol. Sociobiol.*, 58, 270–276.

Engels W., Rozenkranz P., Adler A., Taghizaden T., Luebke G., Franko W. (1997) Mandibular gland volatiles and their ontogenetic patterns in queen honey bees, *Apis meliffera carnica. J. Insect. Phys.*, 43, 307–313.

Free J.B. (1987) *Pheromones of Social Bees*, Cornell University Press, Ithaca, New York.

von Frisch K. (1927) *Aus dem Leben der Bienen.* Springer-Verlag Berlin/Heidelberg/New York (9. Auflage 1977).

von Frisch K. (1965) *Tanzsprache und Orientirung der Bienen*. Springer, Berlin.

Gilley D.C., DeGrandi-Hoffman G., Hooper J.E. (2006) Volatile compounds emitted by live European honey bee (*Apis meliffera* L.) queens. *J. Insect Physiol.* 52, 520–527.

Giurfa M. (1993) The repellent scent-mark of the honeybee *Apis mellifera ligustica* and its role as communication cue during foraging. *Insect. Soc.*, 40, 59–67.

Harbo J.R., Harris J.W. (2005) Suppressed mite reproduction explained by the behaviour of adult bees. *J. Apicul. Res.*, 44, 21–23.

Harris J. (2007) Bees with *Varroa* sensitive hygienic preferentially remove mite-infested pupae aged five days post-capping. *J. Apicul. Res.*, 46, 134–139.

Huang Z.-Y., Robinson G.E. (1992) Honey bee colony integration: Worker-worker interactions mediate plasticity in endocrine and behavioral development. *Proc. Nat. Acad. Sci.*, 89, 11726–11729.

Isidorov V.A., Zenkevich I.G., Ioffe B.V. (1985) Volatile organic compounds in the atmosphere of forests. *Atmos. Environ.*, 19, 1–13.

Kablau A., Berg S., Härtel S., Schneider R (2020a) Hyperthermia treatment can kill immature and adult *Varroa destructor* mites without reducing drone fertility. *Apidologie*, 52, 307–315.

Kablau A., Berg S., Rutschmann B., Scheiner R. (2020b) Short-term hyperthermia at larval age reduces sucrose responsiveness of adult honeybees and can increase life span. *Apidologie*, 52, 570–582.

Kaissling K.E., Renner M. (1968) Antennale Rezeptoren für Queen Substance und Sterzelduft bei der Honigbiene. *Z. Vergl. Physiol.* 59, 357–361.

Kaminski L.-A., Slessor K.N., Winston M.L., Hay N.W., Borden J.N. (1990) Honey bee response to queen mandibular pheromone in laboratory bioassays. *J. Chem. Ecol.* 16, 841–850.

Karpov B., Zabelin B. (1978) Heat treatment for the control of *Varroa jacobsoni* infestation in bees. Veterenaria, № 5, 121–122.

Kasperek K., Paleolog J. (2011) Can spraying bees during winter be an adaptive defense mechanism of colonies against a nosemosis? In: XLVII Beekeep. Sci. Conf. Pszczyna, April 5-7, 2011, pp. 78-79 (in Polish).

Katzav-Gozansky T., Soroker V., Hefetz A. (2002) Honeybees Dufour's gland – idiosyncrasy of new queen signal. *Apidoligie*, 33, 525–537.

Khrust I.I. (1978) Thermal treatment during Varroatosis. *Beekeeping*, №6, 5–6 (in Russian).

Keeling C.I., Slessor K.N., Hugo H.A., Winston M.L. (2003) New components of honey bee (*Apis mellifera* L.) queen retinue pheromone. *Proc. Natl. Acad. Sci. USA*, 100, 14486–14491.

Koeniger N., Koeniger G. (2000) Reproductive isolation among species of the genus *Apis*. *Apidologie*, 31, 313–339.

Koeniger N., Koeniger G., Gries M., Tingek S. (2005) Drone competition at drone congregation areas in four *Apis* species. *Apidology*, 36, 211–221.

LeConte Y., Arnold G., Trouiller J., Masson C. (1990) Identification of a brood pheromone in honeybees. *Naturwissenschaften*, 81, 462–465.

LeConte Y., Sreng L., Trouiller J., Poitou S.H. (1997) Process for modulating the behavior of worker bees by means of brood pheromones. Patent USA 5695383.

LeConte Y., Bécard J.-M., Costagliola G., de Vaublanc G., El Maâtaoui M., Crauser D., Plettner E., Slessor K.N. (2006) Larval salivary glands are a source of primer and releaser pheromone in honey bee (*Apis mellifera* L.). *Naturwissensch.* 93, 237–241.

Lem S. (1989) Parasite strategies, the AIDS virus, and one evolutionary hypothesis. *Priroda*, № 5, 96–104 (in Russian).

Lensky Y., Cassier P., Notkin M., Delorme-Joulie C., Levinsohn M. (1985)

Pheromonal activity and fine structure of the mandibular glands of honeybee drones (*Apis mellifera*) (Insecta, Hymenoptera, Apidae). *J. Insect Physiol.*, 31, 265–276.

Leoncini I., Le Conte Y., Costagliola G., Plettner E., Toth A.L., Wang M., Huang Z., Bécard J.-M., Crauser D., Slessor K.N., Robinson G.E. (2004) Regulation of behavioral maturation by a primer pheromone produced by adult worker honey bees. *Proc. Nat. Acad. Sci. USA*, 101, 17559–17564.

Locke B. (2016) Natural *Varroa* mite-surviving *Apis mellifera* honeybee populations. *Apidologie*, 47, 467–482.

Locke B., Fries I. (2011) Characteristics of honey bee colonies (*Apis mellifera*) in Sweden surviving *Varroa destructor* infestation. *Apidologie*, 42 (4), 533 542.

Malka O., Shnieor S., Hefetz A., Katzav-Gozansky T. (2007) Reversible royalty in worker honeybees (*Apis mellifera*) under the queen influence. *Behav. Ecol. Sociobiol.*, 61, 465–473.

Martin S., Beekman M., Wossler T.C., Ratnieks F.L.W. (2002) Self-replicating honeybees evade worker policing. *Nature*, 415, 163–165.

Mohammedi A., Crsauser D., Paris A., Le Conte Y. (1996) Effect of a brood pheromone on honeybee hypopharyngeal glands. *C.R. Acad. Sci. Paris, Sci. Vie.*, 319, 769–772.

Mohammedi A., Crsauser D., Le Conte Y. (1998) Effect of aliphatic esters on ovary development of queenless bees (*Apis mellifera* L.). *Naturwissenschaften*, 85, 455–458.

Moritz R.F.A., Crewe R.M. (1991) The volatile emissions of honeybee queens (*Apis mellifera* L.). *Apidologie*, 22, 205–212.

Moritz R.F.A., Fuchs S. (1998) Organization of honeybee colonies: characteristics and consequences of a superorganism concept. *Apidologie*, 29, 7–21.

Moritz R.F.A., Lattorff H.M.G., Crewe R.M. (2003a) Honeybee workers (*Apis mellifera capensis*) compete for producing queen-like pheromone signals. *Proc. Roy. Soc. Lond.* B (Suppl.). DOI 10.1098/rsbl.2003.0113.

Moritz R.F.A., Lattorff H.M.G., Crewe R.M. (2003b) Lethal fighting between honeybee queens and parasitic workers (*Apis mellifera*). *Naturwissenschaften*, 90, 378–381.

Oldroyd B.P., Smolenski A.J., Cornuet J.M., Crozier R.H. (1994) Anarchy in the beehive. *Nature*, 371, 749.

Page R.E., Jr., Pankiw T. (2003) Synthetic bee pollen foraging pheromone and uses thereof. Patent USA, 6595828.

Pankiw T., Page R.E., Jr. (2001) Brood pheromone modulates honeybee (*Apis mellifera* L.) sucrose response thresholds. *Bechav. Ecol. Sociobiol.*, 49, 206–213.

Pankiw T. (2004) Cued in: honey bee pheromones as information flow and collective decision-making. *Apidologie*, 35, 217–226.

Peters L., Zhu-Sazman K., Pankiw T. (2010) Effect of primar pheromones and pollen diet on the food producing glands of worker honey bees (*Apis mellifera* L.). *J. Insect Phys.*, 56, 132–137.

Ratnieks F.L.W., Visscher P.K. (1989) Worker policing in the honey bee. *Nature*, 342, 796–797.

Seeley T.D. (1989) The honey bee colony as a superorganism. *Am. Sci.*, 77, 546–553.

Seeley Th.D., Tarpy D.R., Griffin S.R., Carcione A., Delaney D.A. (2015) A survivor population of wild colonies of European honeybees in the northeastern United States: in vestigating its genetic structure. *Apidologie*, 46, 654–666.

Slessor K.N., Kaminski L.A., King G.G.S., Borden J.H., Winston M.L. (1988) Semichemical basis of the retinue response to queen honey bees. *Nature*, 332, 354–356.

Slessor K.N., Kaminski L.A., King G.G.S., Gaylord G.S., Borden J.H., Winston M.L. (1991) Novel pheromone composition for use in controlling honey bee colonies. Patent USA 4990331.

Schmidt J.O., Morgan E.D., Oldham N.J., Do Nascimento R.R., Dani F.R. (1997) (Z)-11-eicosen-1-ol, a major component of *Apis cerana* venom. *J. Chem. Ecol.*, 8, 1923–1939.

Schmidt J.O. (1999) Attractant or pheromone: the case of Nasonov secretion and honeybee swarms. *J. Chem. Ecol.*, 25, 2051–2056.

Schmitt T., Herzner G., Weckerle B., Schreier P., Strohm E. (2007) Volatiles of foraging honeybees *Apis mellifera* (Hymenoptera: Apidae) and their potential role as semichemicals. *Apidologie*, 38, 164–170.

Stout J.C., Goulson D. (2001) The use of conspecific and interspecific scent marks by foraging bumblebees and honeybees. *Animal Behav.*, 62, 183–189.

Tarpy D.R., Gilley D.C. (2004) Group decision making during queen production in colonies of highly eusocial bees. *Apidologie*, 35, 207–216.

Tarpy D.R., Seeley T.D. (2006) Lower disease infection in honeybee (*Apis mellifera*) colonies headed by polyandrous vs monoandrous queens. *Naturwissenschaften*, 93, 195–199.

Tautz J. (2008) *Phänomen Honigbiene*. 2006, Springer-Verlag, Berlin Heidelberg.

Vallet A., Cassier P., Lensky Y. (1991) Ontogeny of the fine structure of the mandibular glands of the honeybee (*Apis mellifera* L.) workers and the pheromonal activity of 2-heptanone. *J. Insect Physiol.*, 37, 789–804.

Wanner K.W., Nichols A.S., Walden K.K.O., Brockmann A., Luetje C.W., Robertson H.M. (2007) A honey bee odorant receptor for the queen substance 9-oxo-2-decenoic acid. Proc. Nat. Acad. Sci. USA, 104, 14383–14388.

Watmough J., Winston M.L., Slessor K.N. (1998) Distinguishing the effects of dilution and restricted movement on the intra-nest transmission of honey-bee queen pheromones. *Bull. Math. Biol.*, 60, 307–327.

Woyke J. (1998) Porozumiewanie się pszczół. W: *Pszczelnictwo*. Pod red. J. Prabuckiego. Wyd. „Albatros", Szczecin, s. 224–227.

Chapter 3

Barker S.A., Foster A.B., LambD.C. (1959) Identification of 10-hydroxy-2-decenoic acid in royal jelly. *Nature*, 183, 996–997.

Beetsma J.(1979) Process of queen–worker differentiation in the honeybee. *Bee World*, 60, 24–39.

Blum M.S., Novak A.F., Taber III, S. (1959) 10-Hydroxy-2-decenoic acid, an antibiotic found in royal jelly. *Science*, 130, 452–453.

Boch R., Shearer D.A. (1967) 2-Heptanone and 10-hydroxy-*trans*-dec-2-enoic acid in the mandibular glands of worker honey bees of different ages. *Zeitschr. Vergl. Physiol.*, 54, 1–11.

Boch R., Shearer D.A., Shuel R.W. (1979) Octanoic acid and other volatile acids in the mandibular glands of the honeybee and in royal jelly. *J. Apicul. Res.*, 18, 250–252.

Boselli E., Caboni M.F., Sabatini A.G., Marcazzan G.L., Lercker G. (2003) Determination and changes of free amino acids in royal jelly during storage. *Apidologie*, 34, 129–137.

Boukraa L., Niar A., Benbarek H., Benhanifia M. (2008) Additive action of royal jelly and honey against *Staphylococcus aureus*. *J. Med. Food*, 11, 190–192.

Calderone N.W., Lin S., Kuenen L.P.S. (2002) Differential infestation of honeybee, *Apis mellifera*, worker and queen brood by the parasitic mite *Varroa destructor*. *Apidologie*, 33, 389–398.

Cymborowski B. (1998) Fizjologia pszczoły. W: *Pszczelnictwo.* Pod red. J. Prabuckiego. Wyd. "Albatros", Szczecin, 1998, s. 173–179.

Dixit P.K., Patel N.G. (1964) Insulin-like activity in larval foods of the honeybee. *Nature*, 202, 189–190.

Drijfhout F.P., Kochansky J., Lin S., Calderone N.W. (2005) Components of honeybee royal jelly as deterrents of the parasitic *Varroa* mite, *Varroa destructor*. *J. Chem. Ecol.*, 31, 1747–1764.

Dzopalic T., Vucevic D., Tomic S., Djokic J., Chinou I., Colic M. (2011) 3,10-Dihydroxy-decanoic acid, isolated from royal jelly, stimulated Th1 polarising capability of human monocyte-derived dendritic cells. *Food Chem.*, 126, 1211–1217.

Engels W., Rosenkranz P., Adler A., Taghizadeh T., Lübke G., Francke W. (1997) Mandibular gland volatiles and their ontogenetic patterns in queen honey bees, *Apis mellifera carnica*. *J. Insect Physiol.*, 43, 307–313.

Fujii A., Kobayashi S., Kuboyama N., Furukawa Y., Kaneko Y., Ishihama S. Yamamoto H., Tamura T. (1990) Augmentation of wound healing by royal jelly (RJ) in streptozotocin-diabetic rats. *Jap. J. Pharmacol.*, 53, 331–337.

Fontana R., Mendes M.A., de Souza B.M., Konno K., César L.M., Malaspina O., Palma M.S. (2004). Jelleines: a family of antimicrobial peptides from royal jelly of honeybees (*Apis mellifera*). *Peptides*, 25, 919–928.

Fujiwara S., Imai J., Fujiwara M., Yaeshima T., Kawashima T., Kobayashi K. (1990) A potent antibacterial protein in royal jelly. Purification and determination of the primary structure of royalisin. *J. Biol. Chem.*, 265, 11333–11337.

Goras G., Tananaki C., Gounari S., et al. (2015) Hyperthermia – a non-chemical control strategy against varroa. J. Hell. Vet. Med. Soc., 66 (4), 249–256.

Hashimoto M., Kanda M., Ikeno K., Hayashi Y., Nakamura T., Ogawa Y., Fukumitsu H., Nomoto H., Fukurawa S. (2005) Oral administration of royal jelly facilitates mRNA expression of glial cell line-derived neurotrophic factor and neurofilament H in the hippocampus of the adult mouse brain. *Biosci. Biotechnol. Biochem.* 69, 800–805.

Hattori N., Nomoto H., Fukumitsu H., Mishima S., Furukawa S. (2007) Royal jelly and its unique fatty acid, 10-hydroxy-*trans*-2-decenoic acid, promote neurogenesis by neural stem/progenitot cells in vitro. *Biomed. Res.* 28, 261–266.

Iizuka T., Iwadare T., Orito K. (1979) Antibacterial activity of myrmicacin and related compounds on pathogenic bacteria in silkworm larvae, *Streptococcus faecalis* AD-4. *J. Fac. Agr. Hokkaido Univ.*, 59, 262–266.

Inoue S.I., Koya-Miyata S., Ushio S., Iwak K., Ikeda M., Kurimoto M. (2003) Royal jelly prolongs the life span of C3H/H3J mice: correlation with reduced DNA damage. *Exp. Gerontol.*, 38, 965–969.

Isidorov V.A., Czyżewska U., Isidorova A.G., Bakier S. (2009) Gas chromatographic and mass spectrometric characterization of the organic acids extracted from some preparations containing lyophilized royal jelly. *J. Chromatogr.* B, 877, 3776–3780.

Isidorov V., Bakier S. (2011) Tajemnice mleczka pszczelego. *Pszczelarstwo*, № 5, 2–4.

Isidorov V.A., Bakier S., Grzech I. (2012) Gas chromatographic–mass spectrometric investigation of volatile and extractive compounds of crude royal jelly. *J. Chromatogr.* B, 885-886, 109–116.

Istomina-TzvetkovaK.P. (1953) Reciprocal feeding of bees. *Pchelovodstvo*, 30, 25–29.

Izuta H., Chikaraishi Y., Shimazawa M., Mishima S., Hara H. (2007) 10-Hydroxy-2-decenoic acid, a major fatty acid from royal jelly, inhibits VEGF-induced angiogenesis in human umbilical vein endothelial cells. *Biosci. Biotechnol. Biochem.*, 71, 253–255.

Jamnik P., Goranovič D., Raspor P. (2007) Antioxidative action of royal jelly in the yeast cell. *Exper. Gerontol.*, 42, 594–600.

Kanbur M., Eraslan G., Beyaz L., Silici S., Liman B.C., Altinordulu S., Atasever A. (2009) The effects of royal jelly on liver damage induced by paracetamol in mice. *Exp. Toxicol. Pathol.*, 61, 123–132.

Karaali A., Meydanoglu F., Eke D. (1988) Studies on composition, freeze-drying and storage of Turkish royal jelly. *J. Apicul. Res.*, 27, 182–185.

Kerr W.E., Blum M.S., Pisani J.F., Stort A.C. (1974) Correlation between amounts of 2-heptanone and iso-amyl acetate in honeybees and their aggressive behaviour. *J. Apicul. Res.*, 13, 173–176.

Kimura Y., Takaku T., Okuda H. (2003) Antitumor and antimetastatic actions by royal jelly in Lewis lang carcinoma-bearing mice. *J. Tradit. Med.*, 20, 195–200.

Kramer, K.J., Tager, H.S., Childs, C.N., Speirs, R.D. (1977) Insulin-like hypoglycemic and immunological activities in honeybee royal jelly. *J. Insect Physiol.* 23, 293–295.

Lem S. (2000) Okamgnienie. Wyd. Literackie, Kraków.

Lercker G., Capella P., Conte L.S., Ruini F., Giordani G. (1981) Components of royal jelly. I. Identification of organic acids. *Lipids,* 16, 912–919.

Lercker G., Savioli S., Vecchi M.A., Sabatini A.G., Nanetti A., Piana L. (1986) Carbohydrate determination of royal jelly by high resolution gas chromatography (HRGC). *Food Chem.*, 19, 255–264.

Liming W., JinhuiZh., Xiaofeng X., Yi L., Jing Zh. (2009) Fast determination of 26 amino acids and their content changes in royal jelly during storage using ultra-performance liquid chromatography. *J. Food Comp. Anal.*, 22, 242–249.

McCleskey C.S., Melampy R.M. (1938) *J. Bacteriol.*, 36, 324 [цит. по: Fujiwara et al., 1990].

Melliou E., ChinouI. (2005) Chemistry and bioactivity of royal jelly from Greece. *Agric. Food Chem.*, 53, 8987–8992.

Mishima S., Suzuki K.-M., Isohama Y., Kuratsu N., Araki Y., Inoue M., Miyata T. (2005) Royal jelly has estrogenic effects in vitro and in vivo. *J. Ethnopharm.*, 101, 215–220.

Nagai T., Sakai M., Inoue R., Inoue H., Suzuki N. (2001) Antioxidative activities of some commercially honeys, royal jelly, and propolis. *Food Chem.*, 75, 237–240.

Nagai T., Inoue R. (2004) Preparation and the functional properties of water extract and alkaline extract of royal jelly. Antioxidant effect, royalisin – antibacterial protein. *Food Chem.*, 84, 181–186.

Nagai T., Sakai M., Inoue R., Inoue H., Suzuki N. (2006) Antioxidant properties of enzymatic hydrolysates from royal jelly. *J. Med.Food.*, 9, 363–367.

Naik D.G., Katke S., Chawda S.S., Thomas D. (1997) 2-Heptanone as a repellent for *Apis cerana*. *J. Apicul. Res.*, 36, 151–154.

Naik D.G., Banhatt P., Chadawa S.S., Thomas D. (2002) 2-Heptanone as a repellent for *Apis florae*. *J. Apicul. Res.*, 40, 59–61.

Nakaya M., Onda H., Sasaki K., Yukiyoshi A., Toshibana H., Yamada K. (2007). Effect of royal jelly

on bisphenol A-induced proliferation of human breast cancer cells. *Biosci. Biotechnol. Biochem.*,

71, 253–255.

Nazzi F., Bortolomeazzi R., Della Vedova G., Del Piccolo F., Annoscia D., Milani N. (2009) Octanoic acid confers to royal jelly varroa-repellent properties. *Naturwissenschaften*, 96, 309–314.

Patel N.G., Haydak M.H., Gochnauer T.A. (1960) Electrophoretic components of the proteins in honeybee larval food. *Nature*, 186, 633–634.

Peixoto L.G., Calábria L.K., Garcia L., Capparelli F.E., Goulart L.R., de Sousa M.V., Espindola F.S. (2009) Identification of major royal jelly proteins in the brain of the honeybee *Apis mellifera*. *J. Insect Physiol.*, 55, 671–677.

Peters L., Zhu-Zalzman K., Pankiw T. (2010) Effect of primer pheromones and pollen diet on the food producing glands of worker honey bees (*Apis mellifera* L.). *J. Insect Physiol.*, 56, 132–137.

Pletner E., Slessor K.N., Winston M.L., Oliver J.E. (1996) Caste-selective pheromone biosynthesis in honeybees. *Science*, 271, 1851–1853.

Sabatini A.G., Marcazzan G.L., Caboni M.F., Bogdanov S., de Almeida-Muradian L.B. (2009) Quality and standardization of royal jelly. *J. ApiProd. ApiMed. Sci.*, 1, 16–21.

Santos K.S., dos Santos L.D., Mendes M.A., de Souza B.M., Malaspina O., Palma M.S. (2005) Profiling the proteome complement of the secretion from hypopharyngeal glands of Africanized nurce-honeybees. *Insect Biochem. Mol. Biol.*, 35, 85–91.

Schildknecht H., Koob K(1971) The first insect herbicide. *Angew. Chem. Int. Ed.*, 10, 124 – 125.

Schmitzová J., Klaudiny J., Albert Š., Schröder W., Schreckengost W., Hanes J., Júdová J. (1998). A family of major royal jelly proteins of the honeybee *Apis mellifera* L. *Cell. Mol. Life Sci.*, 54, 1020–1030.

Schönleben S., Sickmann A., Mueller M.J., Reinders J. (2007) Proteome analysis of *Apis mellifera* royal jelly. *Anal. Bioanal. Chem.*, 389, 1087–1093.

Sesta G. (2006) Determination of sugars in royal jelly by HPLC. *Apidologie*, 37, 84–90.

Stamenković-Radak M., Savić T., Vićentić M., Anđelković M. (2005) Antigenotoxic effects of royal jelly in the sex linked recessive lethal test with *Drosophilla melanogaster*. *Acta Veter.*, 55, 301–306.

Steyn D.G. (1973) Molecular structure and functions of food carbohydrates. *Ind. Univ. Coop. Symp.*, 81–107.

Suemaru K., Cui R., Li B., Watanabe S., Okihara K., Hashimoto K., Yamada H., Araki H. (2008) Topical application of royal jelly has a healing effect for 5-fluorouracil-induced experimental oral mucositis in hamsters. *Methods Find. Exp. Clin. Pharmacol.*, 30, 103.

Timins J.K.(2004) Current issues in hormone replacement therapy.*New Jersey Med.*, 101, 21–27.

Townsed G.F., Lucas C.C. (1940) Chemical nature of royal jelly. *Biochem. J.*, 34, 1155–1162.

Townsed G.F., Morgan J.F., Hazlett B. (1959) Activity of 10-hydroxydecenoic acid from royal jelly against experimental leukemia and ascetic tumors. *Nature*, 183, 1270–1271.

Townsed G.F., Morgan J.F., Tolnai S., Hazlett B., Morton H.J., Shuel R.W. (1960) Studies on the in vitro antitumor activity of fatty acid I. 10-Hydroxy-2-decenoic acid from royal jelly. *Cancer Res.*, 20, 503–510.

Townsed G.F., Brown W.H., Felauer E.E., Hazlett B. (1961) Studies on the in vitro antitumor activity of fatty acid. IV. The esters of acids closely related to 10-hydroxy-2-decenoic acid from royal jelly against plantable mouse leukaemia. *Can. J. Biochem. Physiol.*, 39, 1765–1770.

Vergoz V.,Schreus H.A., Mercer A.R.(2007) Queen pheromone blocks aversive learning in young worker bees. *Science*, 317, 384–386.

Vucevic D., Melliou E., Vasilijic S., Gasic S., Ivanovski P., ChiouI., Colic M. (2007) Faty acid isolated from royal jelly modulate dendritic cell-mediated immune response in vitro. *Intern. Immunopharm.*, 7, 1211–1220.

Winston M.L. (1991) *The Biology of the Honey Bee*. HarvardUniv.Press, England, s. 67.

Witkowska D., Bartys A., Gamian A. (2008) Defensyny i katelicydyny jako naturalne antybiotyki peptydowe. *Post. Hig. Med. Dosw. (on line)*, 62, 694–707.

Xue X.F., Zhou J.H., Wu L.M., Fu L.H., Zhao J. (2009) HPLC determination of adenosine in royal jelly. *Food Chem.*,doi: 10.1016/j.foodchem.2008.12.003.

Zeng H.-Q., Hu F.L., Dietemann V. (2011) Changes in composition of royal jelly harvested at different times: consequences for quality standards. *Apidologie*, 42, 39–47.

Chapter 4

Isidorov V. (2012). Drone brood homogenate – a valuable but underestimated bee product. Beekeeping, № 3, 2–4 (in Polish).

Kabała-Dzik A., Smagacz O., Marquardt W., Stojko A., Szaflarska-Stojko E., Wyszyńska M. (2007a) Protecting effect of bee brood DNA in relation to embryo toxic compounds - acetylsalicylic acid. In: XLIV Beekeep. Sci. Conf.., Puławy, p. 138–139 (in Polish).

Kabała-Dzik A., Smagacz O., Marquardt W., Stojko A., Szaflarska-Stojko E., Wyszyńska M. (2007b) Pharmacological properties of bee brood. In: XLIV Beekeep. Sci. Conf., Puławy, p. 139–140 (in Polish).

Lazaryan D.S. (2002) Comparative amino acid analysis of bee brood. *Pharm. Chem. J.*, 36, 680–682.

Lazaryan D.S., Sotnikova E.M., Evtushenko N.S. (2003) Standardization of bee brood homogenate composition. *Pharm. Chem. J.*, 37, 614–616.

Li J., Fang Y. Zhang L., Begna D. (2011) Honeybee (*Apis mellifera ligustica*) drone embyo proteomes. *J. Insect Physiol.*, 57, 372–384.

Meda A., Lamien Ch.E., Millogo J., Romito M., Nacoulma O.G. (2004) Therapeutic uses of honey and honeybee larvae in central Burkina Faso. *J. Ethnopharm.*, 95, 103–107.

Pemberton R.W. (1999) Insect and other arthropods used as drugs in Korean traditional medicine. *J. Ethnopharm.*, 65, 207–216.

Stojko A. (2007) Apitherapy - its current state and hopes for the future. In: XLIV Beekeep. Sci. Conf., Puławy, p. 142–144 (in Polish).

VasilenkoYu.K., Klimova I.I., Lazaryan D.S. (2002) Biological effect of drone brood under chronic hyperlipidemia conditions. *Pharm. Chem. J.*, 36, 434–436.

VasilenkoYu.K., Klishina I.I., Lazaryan D.S. (2005) A comparative study of the immunotropic and hepatotropic action of beekeeping products in rats with drug-induced hepatitis . *Pharm. Chem. J.*, 39, 319–321.

Chapter 5

Bankova V., Popova M., Bogdanov S., Sabatini A.-G. (2002) Chemical composition of European propolis: expected and unexpected results. *Z. Naturforsch.*, 57c, 530–533.

Banskota A.H., Nagaoka T., Sumioka L.Y., Tezuka Y., Awale S., Midorikawa K., Matsushige K., Kadota S. (2000) Antiproliferative activity of the Netherland propolis and its active principles in cancer cell lines. *J. Etnopharmacol.*, 80, 67–73.

Başer K.H.C., Demirci B. (2007) Studies of *Betula* essential oils. *Arkivos*, 7, 335–348.

Borčić I., Radonić A., Grzunov K. (1996) Comparison of volatile constituents of propolis gathered in different regions of Croatia. *Flav. Fragr. J.*, 11, 311–313.

Burdock G.A. (1998) Review of the biological properties and toxicity of bee propolis (propolis). *Food Chem. Toxicol.*, 36, 347–363.

Demirci B., Başer K.H.C., Özek T., Demirci F. (2000) Betulenols from *Betula* species. *Planta Med.*, 66, 490–93.

Drago L., Mombelli B., De Vecchi E., Fassina M.C., Tocalli L., Gismondo M.R. (2000) In vitro antimicrobial activity of propolis dry extract. *J. Chemotherapy*, 12, 390–395.

Frenkel K., Wei H., Bhimani R., Ye J., Zadunaisky J.A., Huang M.-T., Ferraro T., Conney A.H., Grunberger D. (1993) Inhibition of tumor promotermediated processes in mouse skin and bovine lens by caffeic acid phenethyl ester. *Cancer Res.*, 53, 1255–1261.

Galashkina N.G., Vedernikov D.N., Roshchin V.I. (2004) Birch bud flavonoids *Betula pendula* Roth. *Rastit. Res.*, 40, 62–68 (in Russian).

Gałuszka H. (1998) Bee anatomy. In: Beekeeping. Edited by. J. Prabuckiego. Publ. „Albatros", Szczecin, pp.147–148 (in Polish).

Glinka Ł. (2008) The study of the chemical composition of biologically active substances in propolis and its plant precursors. Master thesis. Dissertation advisor V. Isidorov, UwB, Białystok.

Greenaway W., Scaysbrook T., Whatley F.R. (1988) Composition of propolis in Oxfordshire, U.K. and its relation to poplar bud exudate. *Z. Naturforsch.*, 43c, 301–305.

Greenaway W., May J., Whatley F.R. (1989) Flavonoid aglycones identified by gas chromatography-mass spectrometry in bud exudate of *Populus balsamifera*. *J. Chromatogr.*, 472, 393–400.

Greenaway W., May J., Scaysbrook T., Whatley F.R. (1990) Identification by gas chromatography-mass spectrometry of 150 compounds in propolis. *Z. Naturforsch.*, 46c, 111–121.

Greenaway W., Whatley F.R. (1990) Resolution of complex mixtures of phenolics in poplar bud exudate by analysis of gas chromatography-mass spectrometry data. *J. Chromatogr.*, 519, 145–158.

Greenaway W., Gümüsdere I., Whatley F.R. (1991) Analysis of phenolics of bud exudate of *Populus euphratica* by GC-MS. *Phytochemistry*, 30, 1883–1885.

Greenaway W., Whatley F.R. (1991) Analysis of phenolics of bud exudate of *Populus ciliata* by GC-MS. *Phytochemistry*, 30, 1887–1889.

Grzech I. (2011) Application of the HS-SPME / GC-MS method to determine the chemical composition of propolis and its plant precursors. Master thesis. Dissertation advisor V. Isidorov, UwB, Białystok.

Gunasekera S.P., Kinghorn A.D., Cordell G.A., Farnsworth N.R. (1981) Plant anticancer agents. XIX Constituents of *Aquilaria malaccensis*. *J. Nat. Prod.*, 44, 569–572.

Hausen B.M., Wollenweber E., Senff H., Post B. (1987a) Propolis allergy I. Origin properties usage and literature review. Propolis allergy II. The sensitizing properties of 1,1-dimethylallyl caffeic acid ester. *Contact Dermat.*, 17, 163–170.

Hausen B.M., Wollenweber E., Senff H., Post B. (1987b) Propolis allergy II. The sensitizing properties of 1,1-dimethylallyl caffeic acid ester. *Contact Dermat.*, 17, 171–177.

Haydak M.H. (1953) *Propolis, Report Iowa State Apiarist*, pp. 74–87.

Havsteen R. (1983) Flavonoids, a class of natural products of high pharmacological potency. *Biochem. Pharmacol.*, 32, 1141–1148.

Huber F. (1814) *Nouvelles Observations sur les Abeilles.* Barde, Magnet and Co., Geneva.

Ioirish N.P. (1976) Beekeeping Products and Their Use. Rosselkhozizdat, Moscow (in Russian).

Isidorov V. A., Brzozowska M., Czyżewska U., Glinka Ł. (2008) Gas chromatographic investigation of phenylpropenoid glycerides from aspen (*Populus tremula* L.) buds. *J. Chromatogr.* A, 1198–1199, 196–201.

Isidorov V. A., Isidorova A.G., Szczepaniak L., Czyżewska U. (2009) Gas chromatographic- mass spectrometric investigation of chemical composition of beebread. *Food Chem.*, 115, 1056–1063.

Isidorov V., Glinka Ł., Grzech J. (2011) Chemical composition of commercial preparations of propolis and its possible plant precursors. In: XLVIII Sci. Beekeep. Conf., Pszczyna, p. 115–116.

Isidorov V., Szczepaniak L., Bakier S. (2014) Rapid GC/MS determination of botanical precursors of Eurasian propolis. *Food Chem.*, 142, 101–110.

Isidorov V., Bagan R., Szczepaniak L., Swiecicka I. (2015) Chemical profile and antimicrobial activity of extractable compounds of *Betula litwinowii* (Betulaceae) buds. *Open Chem.*, 13, 125–137.

Isidorov V.A., Bakier S., Pirożnikow E., Zambrzycka M., Swiecicka I. (2016) Selective behaviour of honeybees in acquiring European propolis plant precursors. *J. Chem. Ecol.*, 42, 475–485.

Isidorov V., Witkowski S., Iwaniuk P. et al. (2018) Royal jelly aliphatic acids contribute to antimicrobial activity of honey. J. Apic. Sci. 62, 111–120.

Isidorov V. (2020) GC-MS of Biologically and Environmentally Significant Organic Compounds. TMS Derivatives. Wiley & Sons Ltd., Hoboken, NJ, USA, 706 p.

Kaczmarek F., Dębowski W.J. (1983) The occurrence of α- and β-amylase in propolis. *Acta Polon. Pharm.*, 40, p. 121 (in Polish).

Kasperek K.,Paleolog J.(2011) Could bees splattering during winter be an adaptive defense mechanism of colonies against nosemosis attack?. In: XLVII Sci. Beekeep. Conf. Pszczyna, 2011, p. 78–79 (in Polish).

Khlgatyan S.V., Berzhets V.M., Khlgatyan E.V. (2008) Propolis: composition, biological properties and allergic activity.Usp. Sovrem. Biol., 128, 77–88 (in Russian).

Kędzia B. (2006) Chemical composition and biological activity of propolis from different regions of the world. *Postępy Fitoterap.*, 23–35 (in Polish).

Kędzia B., Hołderna-Kędzia E. (2009) Propolis in the Treatment of Skin Diseases. Toruń (in Polish).

Kivalkina V.P. (1948) Bactericidal properties of propolis. *Beekeeping*, № 10, 50–51 (in Russian).

Kumazawa Sh., Hamasaka T., Nakayama T. (2004) Antioxidant activity of propolis of various geographic origin. *Food Chem.*, 84, 329–339.

Kühnholz S., Seeley T.D.(1997) The control of water collection in honeybee colonies. *Behav. Ecol. Sociobiol.*, 41, 407–412.

Küstenmacher M. (1911) Propolis. *Ber. Dtsch. Pharm. Ges.*, 21, 65–92.

Lem S. (1989) The parasite strategy, the AIDS virus and one evolutionary theory. *Priroda*, № 5, 96–103 (in Russian).

Marcucci M. C. (1995) Propolis: chemical composition, biological properties and therapeutic activity. *Apidologie*, 26, 83–99.

Meyer W. (1956) "Propolis bees" and their activities. *Bee World*, 37, 25–36.

Melliou E., Chinou I. (2004) Chemical analysis and antimicrobial activity of Greek propolis. *Planta Med.*, 70, 1–5.

Melliou E., Stratis E., Chinou I. (2007) Volatile constituents of propolis from various regions of Greece – Antimicrobial activity. *Food Chem.*, 103, 375–380.

Nakamura J., Seeley T. D. (2006) The functional organization of resin work in honeybee colonies. *Behav. Ecol. Sociobiol.*, 60, 339–349.

Nyeko P., Edwards-Jones G., Day R.K. (2002) Honeybee, *Apis mellifera* (Hymenoptera: Apidae), leaf damage on *Alnus* species in Uganda: a blessing or curse in agroforestry? *Bull. Entomol. Res.*, 92, 405–412.

Pan M.-H., Lai C.S., Ho C.T. (2010) Anti-inflammatory activity of natural dietary flavonoids. *Food Funct.*, 1, 15–31.

Polaczek B. (2011) Beekeeping economy and its direct impact on wintering of bee colonies. In: XLVIII Sci. Beekeep. Conf., Pszczyna, p. 38–39 (in Polish).

Popova M., Silici B., Kaftanoglu O., Bankova V. (2005) Antibacterial activity of Turkish propolis and its qualitative and quantitative chemical composition. *Phytomedicine*, 12, 221–228.

Popova M.P., Bankova V.S., Bogdanov S., Tsvetkova I., Naydenski Ch., Marcazzan G.L., Sabatini A.-G. (2007) Chemical characteristics of poplar type propolis of different geographic origin. *Apidologie*, 38, 306–311.

Popova M.P., Chinou I.B., Marekov I.N., Bankova V.S. (2009) Terpenes with antimicrobial activity from Cretan propolis. *Phytochemistry*, 70, 1262–1271.

Popravko S.A. (1982) Protective Substances of Honey Bees. Publ.: Kolos, Moscow (in Russian).

Popravko S.A., Sokolov I.V., Torgov I.V. (1982) New natural triglycerides. *Chem. Nat. Comp.*, № 2, 169–173 (in Russian).

Rao C.V., Desai D., Simi B., Kulkarni N., Amin S., Reddy B.S. (1993) Inhibitory effect of caffeic acid esters on azoxymethane-induced biochemical changes and aberrant crypt foci formation in rat colon. *Cancer Res.*, 53, 4182–4188.

Rösch G.A. (1927) Beobachtungen an Kittharz sammelnden Bienen (*Apis mellifica* L.). *Biol. Zbl.*, 47, 113–121.

Sforcin J.M. (2007) Propolis and the immune system: a review. *J. Ethnopharm.*, 113, 1–14.

Silici S., Kutluca S. (2005) Chemical composition and antimicrobial activity of propolis collected by three different races of honeybee in the same region. *J. Ethnopharm.*, 99, 69–73.

Szczepaniak L., Isidorov V., Szczepaniak U. (2011) Application of chemometric methods to study the origin of propolis. In: XLVIII Sci. Beekeep. Conf., Pszczyna, p. 128–129 (in Polish).

Uvai K., Osanai Y., Imaizumi T., Kanno S.-i., Takeshita M., Ishikawa M. (2008) Inhibitory effect of the alkyl side chain of caffeic acid analogues on lipopolysaccharide-induced nitric oxide production in RAW264.7 macrophages. *Bioorg. Med. Chem.*, 16, 7795–7803.

Vedernikov D.N., Galashkina N.G., Roshchin V.I. (2007) Esters of the buds of Betula pendula Roth. *Rastit. Res.*, 43, 84–92 (in Russian).

Vedernikov D.N., Roshchin V.I. (2010) Extractive substances of birch buds (Betula pendula Roth.): II. Carbonyl compounds and oxides. Ethers. *J. Bioorg. Chem.*, 36, 899–908 (in Russian).

Volpi N., Bergonzini G. (2006) Analysis of flavonoids from propolis by on-line HPLC-electrospray mass spectrometry. *J. Pharm. Biomed. Anal.*, 42, 354–361.

Wilde J. (1998) Dwarf bee, *Apis florea*. In: *Beekeeping*. Ed. by. J. Prabucki. „Albatros", Szczecin, p. 95–104 (in Polish).

Wilson M.B., Spivak M., Hegeman A.D., et al. (2013) Metabolomics reveals the origin of antimicrobial plant resins collected by honey bees. *PLoS One* 8:e77512.

Żyłowska M., Wyszyńska A., Jagusztyn-Krynicka E.K. (2011) Defensins peptides with antibacterial activity. *Post. Mikrobiol.*, 50, 223–234 (in Polish).

Chapter 6

Adams C., Boult C., Deadman B., et al. (2008) Isolation and characterization of the bioactive fraction of New Zealand manuka (*Leptospermum scoparium*) honey. Carbohydr. Res. 343(4), 651–659.

Aliferis K.A., Tarantilis P.A., Harizanis P., Alissandrakis E. (2010) Botanical discrimination and classification of honey samples applying gas chromatography/ mass spectrometry fingerprinting of headspace volatile compounds. *Food Chem.*, 121, 856–862.

Alissandrakis E., Mantziaras E., Tarantilis P.A., Harizanis P.C., Polissiou M. (2010) Generation of linalool derivatives in an artificial honey produced from bees fed with linalool-enriched sugar syrop. *Eur. Food Res. Technol.*, 231, 21–25.

Alissandrakis E., Tarantilis P.A., Pappas Ch., Harizanis P.C., Polissiou M. (2011) Investigation of organic extractives from unifloral chestnut (*Castanea sativa* L.) and eucalyptus (*Eucalyptus globulus* Labill.) honeys and flowers to identification of botanical marker compounds. *LWT – Food Sci. Technol.*, 44, 1042–1051.

Al-Waili N.S., Salom K., Butler G., Ghamdi A.A. (2011) Honey and microbial infections: A review supporting the use of honey for microbial control. *J. Med. Food*. 14, 1079–1096, doi: 10.1089/jmf.2010.0116.

Atrott J., Henle T. (2009) Methylglyoxal in manuka honey–correlation with antibacterial properties. Czech J. Food Sci. 27, 163–165.

Baker H.G. (1977) Non-sugar chemical constituents of nectar. *Apidologie*, 8, 349–356.

Barbier M. (1976) Introduction a l'Écologie Chimique. Masson, Paris.

Bianchi F., Mangia A., Mattarozzi M., Musci M. (2011) Characterization of the volatile profile of thistle honey using headspace solid-phase microextraction and gas chromatography-mass spectrometry. *Food Chem.*, 129, 1030–1036.

Blatt J., Roces F. (2001) Haemolymph sugar levels in foraging honeybees (*Apis mellifera carnica*): dependence on metabolic rate and in vivo measurement of maximal rates of trehalose synthesis. *J. Exp. Biol.*, 204, 2709–2716.

Bogdanov S. (1997) Nature and origin of the antibacterial substances in honey. *Lebensm.-Wiss. Technol.*, 30, 748–753.

Bonnier G. (1878) *Les Nectaries.*Annales des Sciences Naturales. Botanique, t. 8, 5–212.

Borum E., Gunes E. (2018) Microbiological contamination of honeys from different sources in Turkey. *J. Apicul. Sci.*, 62, 89–96.

Bourkaâ L., Niar A., Benbarek H., Benhanifia M. (2008) Additive action of royal jelly and honey against *Staphylococcus aureus*. *J. Med. Food*, 11, 190–192.

Boukraâ L. (2013) Honey in burn and wound management. In: Honey in Traditional and Modern Medicine. Ed. by L. Boukraâ. Poca Raton, CRC Press, pp. 125–158.

Brudzynski K., Abubaker K., St-Martin L., Castle A. (2011) Re-examination the role of hydrogen peroxide in bacteriostatic and bactericidal activities of honey. *Front. Microbiol.*, 2, article 213, doi: 10.3389/fmicb.2011.00213.

Brudzynski K., Miotto D., Kim L., et al. (2017) Active macromolecules of honey form colloidal particles essential for honey antimicrobial activity and hydrogen peroxide production. *Sci. Rep.*,7:76370, doi: 10.1038/s41598-017-08072-0.

Bučeková M., Majtán J. (2016) The MRjP1 honey glycoprotein does not contribute to the overall antibacterial activity of natural honey. *Eur. Food Res. Technol.*, 242, 625–629.

Cokcetin N.N., Pappalardo M., CampbellL.T. et al. (2016) The antimicrobial activity of Australian *Leptospermum* honey correlates with methylglyoxal levels. *PLoS One*, doi: 10.1371/journal.pone.0167780.

Elsass F. (2017) A sweet solution: The use of medical-grade honey on oral mucositis in the pediatric oncology patient. *J. Wound Ostomy Contin. Nurs.*, 44, S9. Meeting Abstract: CS05.

Fyfe L., Okoro P., Paterson U., et al. (2017) Compositional analysis of Scottish honeys with antibacterial activity against antibiotic-resistant bacteria reveals novel antimicrobial components. *LWT - Food Sci. Technol.*, 79, 52–59, doi:10.1016/j.lwt.2017.01.023.

Gałuszka H., Koteja J., Tworek K. (1996) Bees on the Honeydew. Nowy Sącz, Publ. „ Sądecki Bartnik" (in Polish).

Goyret J., Farina W.M. (2005) Trophallactic chains in honeybees: a quantitative approach of the nectar circulation among workers. *Apidologie*, 36, 595–600.

Guyot Ch., Scheirman V., Collin S. (1999) Floral origin markers of heather honeys: *Calluna vulgaris* and *Erica arborea. Food Chem.*, 64, 3–11.

Guyot-Declerck Ch., Chevance F., Lermusieau G., Collin S. (2000) Optimized extraction procedure for quantifying norisoprenoids in honey and honey food products. *J. Agric. Food Chem.*, 48, 5850–5855.

Hippocrates. On Ulcers. Internet Classic Archive. Transl. by F. Adams at http://classics.mit.edu/Hippocrates/ulcers.5.5.

Horniackova M., Bučeková M., Valachová I., Majtán J. (2017) Effect of gamma radiation on the antibacterial and antibiofilm activity of honeydew honey. *Eur. Food Res. Technol.*, 243, 81–88, doi 10.1007/s00217-016-2725x.

Hołderna-Kędzia E., Kędzia B. (2002) *Variety Honeys and Their Medicinal Meaning. Ed. Farmers' ministry*, Włocławek.

IARC Monographs on the Evaluation of Carcinogenic Risk to Humans (1991) Vol. 51. Coffee, Tea, Mate, Methylxanthines and Methylglyoxal, p.452.

Isidorov V., Kołtowski Z., Grzech I. (2011a) Identification of volatile organic compounds in the nectar headspace by microextraction to the stationary phase and GC-MS. In: XLVIII Sci. Beekeep. Conf., p. 123–124 (in Polish).

Isidorov V.A., Czyżewska U., Jankowska E., Bakier S. (2011b) Determination of royal jelly acids in honey. *Food Chem.*, 124, 387–391.

Isidorov V., Bakier S. (2011) How is honey made? *Beekeeping*, № 4, 2–5 (in Polish).

Isidorov V., Witkowski S., Iwaniuk P. et al. (2018) Royal jelly aliphatic acids contribute to antimicrobial activity of honey. *J. Apic. Sci.*, 62, 111–120.

Isidorov V., Bagan R., Szczepaniak L., Swiecicka I (2015) Chemical profile and antibacterial activity of extractable compounds of *Betula litwinowii* (Betulaceae) buds. *Open Chem.*, 13, 125–137.

Jabłoński B. (2003) *Methodology of the Study of the Abundance of Flower Nectarization and the Assessment of Plant Honey Yield*. Puławy.

Jasicka-Misiak I., Poliwoda A., Dereń M., Kafarski P. (2012) Phenolic compounds and abscisic acid as potential markers for the floral origin of two Polish unifloral honeys. *Food Chem.*, 131, 1149–1156.

Jerković I., Hegić G., Marijanović Z., Bubalo D. (2010) Organic extractivities from *Mentha* spp. honey and the bee-stomach: methyl syringate, vomifoliol, terpenediol I, hotrienol and other compounds. *Molecules*, 15, 2911–2924.

Jerković I., Marijanović Z. (2010) Oak (*Quercusfrainetto*Ten.) Honeydew honey—Approach to screening of volatile organic composition and antioxidantapacity (DPPH and FRAP assay). *Molecules*, 15, 3744–3756.

Kwakman P.H.S., te Velde A.A., de Boer L., et al. (2010) How honey kills bacteria. *The FASEB J.*, 24, 2567–2582.

Kwakman P.H.S., te Velde A.A., de Boer L., et al. (2011) Two major medical honeys have different mechanisms of bactericidal activity. *PLoS One* 6, e17709, doi: 10.1371/journal.pone.0017709.

KwakmanP.H.S., deBoerL., Ruyter-SpiraC.P., Creemers-MolenaarT., HelsperJ.P.F.G., Vandenbroucke-GraulsC.M.J.E., ZaatS.A.J., teVelde A.A. (2011) Medical-grade honey enriched with antimicrobial peptides has enhanced activity against antibiotic-resistant pathogens. *Eur. J. Clin. Microbiol. Infect. Dis.*, 30, 251–257.

Lipiński M. (2010) *Bee Benefits, Pollination and Plant Honey Yield.* PWRiL Sp. z o.o. & Publ. Sądecki Bartnik (in Polish).

Lyapunov Ya.E., Drebezgina E.S., Elovikova E.A., Legotkina G.I. (2011) Pollen analysis of “exotic” honeys in the markets and fairs of Russia or the whole truth about honey. http://24medok.ru.

Midura T.F., Snewden S., Wood R.M., Arnon S.S. (1979) Isolation of *Clostridium botulinum* from honey. *J. Clin. Microbiol.*, 9, 282–283.

Molan P.C., Russell K.M. (1988) Non-peroxide antibacterial activity in some New Zealand honeys. *J. Apicult. Res.*, 2, 762–767.

Molan P.C. (1992) The antibacterial activity of honey. 1. The nature of the antibacterial activity. *Bee World*, 73, 5–28.

Moore O.A., Smith L.A., Campbell F., et al. (2001) Systematic review of the use of honey as a wound dressing. *BMC Compl. Alter. Med.*, 1, 2–7.

Naef R., Jaquier A., Velluz A., Bachofen B. (2004) From linden flower to linden honey – Volatile constituents of linden nectar, the extract of bee-stomach and ripe honey. *Chem. Biodivers.*, 1, 1870–189.

Nepi M., Stpiczyńska M. (2008) The complexity of nectar: secretion and resorption dynamically regulate nectar features. *Z. Naturforsch.*, 95, 177–184.

Noskowicz-Bieroniowa H. (2009) What can honey? Publisher: Emilia, Kraków (in Polish).

Oryan A., Alemzadeh E., Moshiri A. (2016) Biological properties and therapeutic activities of honey in wound healing: A narrative review and meta-analysis. *J. Tiss. Viab.*, 25, 98–118.

Pacini E., Nepi M., Vesprini J.L. (2003) Nectar biodiversity: a short review. *Plant Syst. Evol.*, 238, 7–21.

PersanoOddo L., Piro R. (2004) Main European unifloral honeys descriptive sheets. *Apidologie*, 35 (Suppl. 1), 38–81.

Plettner F., Slessor K.N., Winston M.L., Oliver J.E. (1996) Caste-selective pheromone biosynthesis in honeybees. *Science*, 271, 1851–1853.

Plutowska B., Chmiel T., Dymerski T., Wardencki W. (2011) A headspace solid-phase microextraction method development and its application in the determination of volatiles in honeys by gas chromatography. *Food Chem.*, 126, 1288–1298.

Singaravelan N., Nee'Man G., Inbar M., Izhaki I. (2005) Feeding responces of free-flying honeybees to secondary compounds mimicking floral nectar. *J. Chem. Ecol.*, 31, 2791–2796.

Soria A.C., Sanz J., Martínez-Castro I. (2009) SPME followed GC-MS: a powerful technique for qualitative analysis of honey volatiles. *Eur. Food Res. Technol.*, 228, 579–590.

Szczęsna T., Rybak-Chmielewska H., Waś E., Kachaniuk K., Teper D. (2011) Characteristics of Polish unifloral honeys. I. Rape honey (*Brassica napus*L. var. *oleifera* Metzger). *J. Apucult. Res.*, 55, 111–119.

Tan S.-T., Holland P.T., Wilkins A.L., Molan P.C. (1988) Extractives from New Zealand honeys. White clover, manuka and kanuka unifloral honeys. *J. Agric. Food Chem.*, 36, 453–460.

Tomas-Barberan F.A., Ferreres F., Garcia-Viguera C., Tomas Lorente F. (1993) Flavonoids in honey of different geographical origin. *Z. Lebensm. Unters. Forsch.*, 196, 38–44.

Valachová I., Bučeková M., MajtánJ. (2016). Quantification of bee-derived peptide Defensin-1 in honey by competitive enzyme-linked immunosorbent assay, a new approach in honey quality control. *Czech J. Food Sci.*, 34, 233–243, doi: 10.1722/422/2015-cjfs.

Wainselboim A.J., Roces F., Farina W.M. (2003) Trophallaxis in honeybees *Apis mellifera* (L.), as related to their past experience at the food source. *Anim. Behav.*, 63, 791–795.

Waś E., Rybak-Chmielewska H., Szczęsna T., Kachaniuk K., Teper D. (2011a) Characteristics of Polish unifloral honeys. II. Lime honey (*Tilia* spp.). *J. Apucult. Res.*, 55, 121–128.

Waś E., Rybak-Chmielewska H., Szczęsna T., Kachaniuk K., Teper D. (2011b) Characteristics of Polish unifloral honeys. III. Heather honey (*Calluna vulgaris* L.). *J. Apucult. Res.*, 55, 129–136.

Weston R.J., Mitchell K.R., Allen K.L. (1999) Antibacterial phenolic components of New Zealand manuka honey. *Food Chem.*, 64, 295–301.

Weston R.J., Brocklebank L.K., Lu Y. (2000) Identification and quantitative levels of antibacterial components of some New Zealand honey. *Food Chem.*, 70, 427–435.

Wolski T., Tambor K., Rybak-Chmielewska H., Kędzia B. (2006) Identification of honey volatile components by solid phase microextraction (SPME) and gas chromatography/mass spectrometry (GC/MS). *J. Apicul. Res.*, 50, 115–126.

Yao L., Datta N., Tomás-Barberán F.A., Ferreres F., Martos I., Singanusong R. (2003) Flavonoids, phenolic acids and abscisic acid in Australian and New Zealand *Leptospermum* honeys. *Food Chem.*, 81, 159–168.

Chapter 7

Ioirish N.P. (1976) Beekeeping Products and Their Uses. Moscow, Rosselkhozizdat.

Isidorov V.A., Czyżewska U., Jankowska E., Bakier S. (2011) Determination of royal jelly acids in honey. *Food Chem.*, 124, 387–391.

Isidorov V.A., Grzech I., Bakier S. (2012) Gas chromatographic-mass spectrometric investigation of volatile and extractable compounds of crude rojal jelly. *J. Chromatogr.* B, doi: 10.1016/j.jchromb.2011.12.025.

Pohorecka K., Skubida P. (2002) Assessment of the possibility of administering extracts of medicinal plants to bees to improve their general condition. In: Mat. XXXIX Sci. Beekeep. Conf., Puławy, p. 58–60 (in Polish).

Juszczak L., Socha R., Rożnowski J., Fortuna T., Nalepka K. (2009) Physicochemical properties and quality parameters of herbhoney. *Food Chem.*, 113, 538–542.

Socha R., Juszczak L., Pietrzyk S., Fortuna T. (2009) Antioxidant activity and phenolic composition of herbhoney. *Food Chem.*, 113, 568–574.

Zalewski A. (2017) Chemical composition of some herbhoney and their plant precursors. Master's work. University in Bialystok.

Chapter 8

Bakier S. (2001) Polish beekeeper in the "Wild West". *Beekeeping*, № 7, 13–15.

Baltrušaitytė V., Venskutonis P.R., Čeksterytė V. (2007) Radical scavenging activity of different floral origin honey and beebread phenolic extracts. *Food Chem.*, 101, 502–514.

Čeksterytė V., Kazlauskas S., Račys J. (2006) Composition of flavonoids in Lithuanian honey and beebread. *Biologia*, № 2, 28–33.

Isidorov V.A., Isidorova A.G., Sczczepaniak L., Czyżewska U. (2009) Gas chromatographic-mass spectrometric investigation of the chemical composition of beebread. *Food Chem.*, 115, 1056–1063.

Kaškonienė V., Venskutonis P.R., Čeksterytė V. (2008) Composition of volatile compounds of honey of various floral origin and beebread collected in Lithuania. *Food Chem.*, 111, 988–997.

Kędzia B., Hołderna-Kędzia E. (2010) Beebread. The Most Valuable Bee Product. Ed. by „Humana Divinis", Toruń (in Polish).

Khalifa S.A.M., Elashal M., Kieliszek M. et al., (2020) Recent insights into chemical and pharmacological studies of bee bread (review). *Trends Food Sci. Technol.*, 97, 300–316.

Mărgăoan R., Strant M., Varadi A. et al. (2019) Bee collected pollen and bee bread: bioactive constituents and health benefits. *Antioxidants*, 8, 568; doi: 10.3390/antiox8120568.

Chapter 9

Akimov I.A. (1993) Varroa jacobsoni bee mite. Kiev, Naukova Dumka (in Russian).

Alfonsus A. (1891) Der Feind der Bienenlaus. *Deutsch. Illustr. Bienenzeit.*, 8, 503–506.

Alfonsus A. (1922) An enemy of mites in the beehive. *Bee World*, 4, 2–3.

Aliano N.A., Ellis M.D. (2005) A strategy for using powdered sugar to reduce *Varroa* populations in honey bee colonies. *J. Apicul. Res.*, 44, 54–57.

Al-Ghamdi A., Al-Abbadi A., Abdullah A., et al. (2020) In vitro antagonistic potential of gut bacteria isolated from indigenous honey bee race of Saudi Arabia against *Paenibacillus larvae*. J. Apicul. Res. DOI: 10.1080/00218839.2019.1706912.

Antunez K., Harriet J., Gende L., Maggi M., Eguaras M., Zunino P. (2008) Efficacy of natural propolis extract in control of American foulbrood. *Veter. Microbiol.*, 131, 324–331.

Ariana A., Ebadi R., Tahmasebi G. (2002) Laboratory evaluation of some plant essences to control *Varroa destructor* (Acari: Varroidae). *Exper. Appl. Acarol.*, 27, 319–327.

Arnold G., Masson C., Le Conte Y., Trouiller J., Chappe B., Ourisson G. (1992) Process for combating "varroatosis" by biological means and devices for implementing this process. USA Patent 5135758.

Barry B.C., Verstraten L., Butler F.T., et al. (2018) The use of airborne ultrasound for Varroa destructor mite control in beehives. Proc. Paper, IEEE Iitern. Ultrason. Symp. (IUS), Kobe, Japan.

Bastos E.M.A.F., Simone M., Jorge D.M., Soares A.E.E., Spivak M. (2008) In vitro study of the antimicrobial activity of Brazilian propolis against *Paenibacillus larvae*. *J. Invert. Pathol.*, 97, 273–281.

Beims H. (2018) Charakterisierung der *Paenibacillus larvae* Genotypen ERIC I-V und alternative Therapieverfahren zur Behandlung der Amerikanischen Faulbrut. PhD Thesis, http://uri.gbv.de/document/gvk:ppn:1034123815.

Beims H., Bunk B., Erler S. et al. (2020) Discovery of *Paenibacillus larvae* ERIC V: Phenotypic and genomic comparison to genotypes ERIC I-IV reveal different inventories of virulence factors which correlate with epidemiological prevalences of American Foulbrood. *Int. J. Med. Microbiol.*, 310, https://doi.org/10.1016/j.ijmm.2020.151394.

Beltrán-RamirezO., Alemán-Lazarini L., Salcido-Neyoy M., Hernandes-Garcia H., Fattel-Fazenda S., Arce-Popoca E., Arellanes-Robleod J., Garcia-Roman R., Vazquez-Vazquez P., Sierra-Santoga A., Villa-Trevino S. (2008) Evidence that the anti-carcinogenic effect of caffeic acid phenyl ester in the resistant hepatoccyte model involves modification of cytochrome P450. *Tixicol. Sci.*, 104, 100–106.

Berube C. (1999) Himalayan ceranaid: development assistance to preserve and promote *Apis cerana* beekeeping in Nepal. Part 1 & 2. *Amer. Bee. J.*, 139, 707–710.

Berube C. (1999) Himalayan ceranaid: development assistance to preserve and promote *Apis cerana* beekeeping in Nepal. Part 2. *Amer. Bee. J.*, 139, 784–787.

Bieńkowska M. (2019) What does *Varroa destructor* eat - hemolymph or fat body? *Beekeeping*, №7, 2–5 (in Polish).

Bilikova K., Popova M., Trusheva B., Bankova V. (2013) New anti-*Paenibacillus larvae* substances purified from propolis. *Apidologie*, 44, 278–285.

Boecking O., Drescher W. (1992) The removal response of *Apis mellifera* L. colonies to brood in wax and plastic cells after artificial and natural infestation with *Varroa jacobsoni* Oud. and to freeze-killed brood. *Exp. Appl. Acarol.*, 16, 321–329.

Bogomolov K., Yarankin V. (2011) Collapse of Bee Colonies. Diseases of Bees. Ryazan (in Russian).

Boligon A.A., de Brum T.F., Zandra M., et al. (2013) Antimicrobial activity of *Scutia buxiflia* against the honeybee pathogen *Paenibacillus larvae*. *J. Invertebr. Pathol.*, 112, 105–107.

Boudegga H., Boughalleb N., Barbouche N., Ben Hamouda M.H., El Mahjoub M. (2010) *In vitro* inhibitory actions of some essential oils on *Ascosphaera apis*, a fungus responsible for honey bee chalkbrood. *J. Apicul. Res.*, 46, 236–242.

Büchler R. (2000) Design and success of a German breeding program for *Varroa* tolerance. *Am. Bee J.*, 140, 662–665.

Chandler D., Sunderlaand K.D., Ball B.V., Davidson G. (2001) Perspective biological control agents of *Varroa destructor* n. sp., an important pest of the European honeybee, *Apis mellifera*. *Biocontr. Sci. Technol.*, 11, 429–448.

Chorbiński P. (2012) Defeat Varroa. Edition II, Ed. BEE & HONEY, Kęty (in Polish).

Chorbiński P. (2016) Zwalczanie warrozy bez użycia ciężkiej chemii. W: Pomóżmy Pszczołom – Pszczoły Pomogą Nam. Sądecki Bartnik, Stróże, s. 19–26.

Coffey M.F. (2007) Biotechnical methods in colony management, and the use of Apiguard® and Exomite™ *Apis* for the control of the Varroa mite (*Varroa destructor*) in Irish honey bee (*Apis mellifera*) colonies. *J. Apicul. Res.*, 49, 213–219.

Damiani N, Maggi M.D., Gende L.B., Faverin C., Eguaras M.J., Marcangeli J.A. (2010) Evaluation of the toxicity of a propolis extract on *Varroa destructor* (Acari: Varroidae) and *Apis mellifera* (Hymenoptera: Apidae) *J. Apicul. Res.*, 49, 257–264.

Donovan B.J., Paul F. (2015) Pseudoscorpions: the forgotten beneficials inside beehives and their potential for management for control of Varroa and other arthropod pests. *Bee World*, 86, 83–87.

Eischen F. (1995) Varroa resistance to fluvalinate. Am. Bee J., 135 (12), 815–816

Ellenhorn M.J., Schonwald S., Ordog G., Wasserberger J. (1997). Ellenhorn's Medical Toxicology: Diagnosis and Treatment of Human Poisoning. 2nd ed. Baltimore, MD: Williams and Wilkins, p. 1730.

Erickson E.H., Jr., DeGrandi-Hoffman G., Whitson R.S. (2008) Control of parasitic mites of honey bees. USA Patent 7423068.

Esch H. (1960) Über die Körpertemperaturen und den Wärmehaushalt von *Apis mellifera*. *Zeitschr. Vergl. Physiol.*, 43, 305–335.

Evans J.D., Armsrong T.-N. (2005) Inhibition of the American foulbrood bacterium *Paenibacillus larvae*, by bacteria isolated from honey bee. *J. Apicul. Res.*, 44, 168–171.

Evans J.D., Pettis J.S. (2005) Colony-level impacts of immune responsiveness in honey bees, *Apis mellifera*. *Evolution*, 59, 2270–2274.

Evans J.D., Spivak M. (2010) Socialized medicine: Individual and communal disease barriers in honey bees. *J. Invert.Pathol.*, 103, 562–572.

Evteeva N.I., Chechetkina U.E., Pechkin A.I., Krylov V.N. (2010) The action of propolis on the enterobacteriaceae of bees. *Beekeeping*, № 2, 18–19 (in Russian).

Fagan L.L., Nelson W.R., Meenken E.D. et al. (2012) Varroa management in small bites. *J. Appl. Entomol.*, 136, 473–475.

Fassbinder C., Grodnitzky J., Coats J. (2002) Monoterpenoids as possible control agents for *Varroa destructor*. *J. Apicul. Res.*, 41, 83–88.

Ferrari M.C.F., Favaro R., Mair S., et al. (2020) Application of Metarhizium anisopliae as a potential biological control of *Varroa destructor* in Italy *J. Apicul. Res.*, 59(4), 528–538.

Fombong A.T., Cham D.T., Nkoba K. et al. (2016) Occurrence of the pseudoscorpions *Ellingsenius ugandanus* and *Paratemnoides pallidus* in honey bee colonies in Cameeroon. *J. Apicul. Res.*, 55, 247–250.

Gal H., Slabezki Y., Lensky Y. (1992) A preliminary report on the effect of Origanum oil and thymol application in honeybee (*Apis mellifera* L.) colonies in a subtropical climate on population level of *Varroa jakobsonii*. *Bee Sci.*, 2, 175–180.

Garedew A., Schmolz E., Schricker B., Lamprecht I. (2002) Microcalorimetric investigation of the action of propolis on *Varroa destructor* mites. *Thermochim. Acta*, 382, 211–220.

Garedew A., Schmolz E., Lamprecht I. (2003) Microcalorimetric and respirometric investigation of the effect of temperature on the antivarroa action of the natural bee product-propolis. *Thermochim. Acta*, 399, 171–180.

Garedew A., Schmolz E., Lamprecht I. (2004) Effect of the bee glue (propolis) on the calorimetrically measured metabolic rate and metamorphosis of the great wax moth *Galleria mellonella. Thermochim. Acta*, 413, 63–72.

Gashout H.A., Guzmán-Novoa E. (2009) Acute toxicity of essential oils and other natural compounds to the parasitic mite, *Varroa destructor*, and to larval and adult worker honey bees (*Apis mellifera* L.). *J. Apicul. Res.*, 48, 263–269.

Genersch E. (2010) American Foulbrood in honey bees and its causative agent, *Paenibacillus larvae. J. Invertebr. Pathol.*, 103, S10–S19.

Gilliam M. (1997) Identification and roles of non-pathogenic microflora associated with honey bees. *FEMS Microbiol. Lett.*, 155, 1–10.

González G., Hinojo M.J., Mateo R., Medina A., Jiménez M. (2005) Occurrence of mycotoxins producing fungi in bee pollen. *Int. J. Food Microbiol.*, 105, 1–9.

Gonzalez V.H. Mantilla B., Mahnert V. (2008) A new host record for *Dasychernes inquilinus* (Arachinida, Pseudoscorpiones, Chernetidae), with an overview of pseudoscorpion–bee relationships. *The J. Arachnol.*, 35, 470–474.

Hallmann C.A., Sorg M., Jongejans E. et al. (2017) More than 75 percent decline over 27 years in total flying insects biomass in protected areass. *PLoS ONE*, 12(10):e0185809, doi: 10.1371/journal.pone.0185809.

Harbo J.R., Harris J.W. (2005) Suppressed mite reproduction explained by the behaviour of adult bees. *J. Apicul. Res.*, 44, 21–23.

Harris J. (2007) Bees with *Varroa* sensitive hygienic preferentially remove mite-infested pupae aged ≤five days post-capping. *J. Apicul. Res.*, 46, 134–139.

Hernandez-Torres H., Garcia-Martinez O., Romero-Napoles J. et al., (2018) Sap beetles (1) of Coahuila, Mexico and effective collecting attractants. *Southwest. Entomol.*, 43(1) 151–166.

Hoyt M. (1965) *The World of Bees.* Coward McCann, Inc., New York.

Howis M., Nowakowski P. (2009) *Varroa destructor* removal efficiency using Beevital Hive Clean preparation. *J. Apucul. Sci*,. 53, 15–20.

Imdorf A., Bogdanov S., Kilchenmann V., Maquelin C. (1995) Apilaife Var: A new varroacide with thymol as the main ingredient. *Bee World*, 76, 77–83.

Isidorov V., Buczek K., Zambrowski G., Miastkowski K., Swiecicka I. (2017) In vitro study of the antimicrobial activity of European propolis against *Paenibacillus larvae*. Apidologie, 48, 411–422.

Isidorov V., Witkowski S., Iwaniuk P., Zambrzycka M., Swiecicka I. (2018a) Royal jelly aliphatic acid contribute to antimicrobial activity of honey. *J. Apicul. Sci.*, 62(2), 111–120.

Isidorov V., Buczek K., Segiet A., Zambrowski G., Swiecicka I. (2018b) Activity of selected plant extracts against honey bee pathogen *Paenibacillus larvae*. Apidologie, 49, 687–704.

Kačaniová M., Kasper J., Terentjeva M. (2019) Antagonistic effect of gut microbiota of honeybee *(Apis mellifera)* against causative agent of American Foulbrood *Paenibacillus larvae*. J. Microbiol. Biotechnol. Food Sci., 9, 478–481.

Khmara P.Ya. (1988) Biological characteristics of honey bees, causing the spread of infectious diseases: Lecture. Kiev (in Ukrainian)

Kistner D.H. (1982) The social insects' bestiary. In: Hermann H.D. (Ed.) Social Insects. Vol. 3, Academic Press, New York, USA, pp. 2–244.

Kraus B., Koeniger N., Fuchs S. (1994) Screening of substances for their effect on *Varroa jacobsonii*: attractiveness, repellency, toxicity and masking effects of ethereal oils. *J. Apicul. Res.*, 33, 34–43.

Kujumgiev A., Tsvetkova I., Serkedjieva Yu., et al., (1999) Antibacterial, antifungal and antiviral activity of propolis of different geographic origin. *J. Ethnophormacol.*, 64, 235–240.

Lamei S., Stephan J.G., Riesbeck K., et al. (2019) The secretome of honey bee-specific lactic acid bacteria inhibits *Paenibacillus larvae* growth. *J. Apicul. Res.*, 58(3),405–412.

Lear E. (2000) Some recent beekeeping observations. *South. Beekeep. Newslett.*, 2, 1–4.

Lipiński Z., Szubstarska D., Szubstarski J. (2011) Sensitivity of the domestic *Varroa* destructor population to pyrethroids (tau-fluvalinate). In: XLVII Scientific Beekeep. Conf., Pszczyna, April 5–7, 2011, p. 69–70 (in Polish).

Lodesani M., Costa C. (2008) Maximizing the efficacy of a thymol based product against the mite *Varroa destructor* by increasing the air space in the hive. *J. Apicul. Res.*, 47, 113–117.

Neuman P., Carreck N.L. (2010) Honey bee colony loss. *J. Apicul. Res.*, 49, 1–6.

Neumann P., Pettis J.S., Schäfer M.O. (2016) Quo vadis *Aethine tumida*? Biology and control of small hive beetle. *Apidologie*, 47, 427–466.

Niu G., Johnson R.M., Berenbaum R.M. (2011) Toxicity of mycotoxins to honeybees and its amelioration by propolis. *Apidologie*, 42, 79–87.

Palmeri V., Campolo O., Zappalà (2007) Evaluation of two methods for applying Apiguard® in an area with continuous nectar flow and brood rearing. *J. Apicul. Res.*, 46, 105–109.

Parish J.B., Scott E.S., Correll R., Hogendoorn K. (2019) Survival and probability of transmission of plant pathogenic fungi through the digestive tract of honey bee workers. *Apidologie*, 50, 871–880.

Polaczek B. (2001) Reakcja samic *Varroa destructor* na larwy trutni diploidalnych w rodzinach pszczelich. *Pszczelarstwo*, Nr. 4, 6.

Popravko S.A. (1982) Protective Substances of Honey Bees. Publ.: Kolos, Moscow (in Russian).

Pusceddu M., Piluzza G., Theodorou P. et al. (2019) Resin foraging dynamics in *Varroa destructor*-infested hives: a case of medication of kin? *Insects Sci.*, 26, 297–310.

Ramos O. Y., Basualdo M., Libonatti C., et al.(2019) Current status and application of lactic acid bacteria in animal production systems with a focus on bacteria from honey bee colonies. *J. Appl. Microbiol.* 128(5), 1248–1260.

Ramsey S.D. Ochoa R., Bauchan G. et al. (2019) *Varroa destructor* feeds primarily on honey bee fat body tissue and not hemolymph. *PNAS*, 116, 249–266.

Read S., Howlett B.G., Donovan B.J. et al. (2014) Culturing chelifers (Pseudoscorpions) that consume Varroa mites. *J. Appl. Entomol.*, 138, 260–266.

Reyes-Escobar O., Dosal-Alonso E., Lara-Alvarez C., et al. (2015) Lethal effect of boric acid and attractants against the small hive beetle, *Aethina tumida* Murray (Coleoptera: Nitidulidae)

J. Apicul. Res., 54(3), 226–232.

Reynaldi F.J., De Giusti M.R., Alippi A.M. (2004) Inhibition of the growth of *Ascosphaera apis* by *Bacillus* and *Paenibacillus* strains isolated from honey. *Rev. Argent. Microbiol.*, 36, 52–55.

Rümmelin K (2004) A method for controlling pests in bee colonies. Patent DE 10161677 B4.

Sabate D.C., Carrillo L., Carina Audisio M. (2009) Inhibition of *Paenibacillus larvae* and *Ascosphaera apis* by *Bacillus subtillis* isolated from honeybee gut and honey samples. *Res. Microbiol.*, 160, 193–199.

Sammataro D., Finley J., LeBlanc B., Wardell G., Fabiana Ahumada-Segura F., Carroll M.J. (2009) Feeding essential oils and 2-heptanone in sugar syrup and liquid protein diets to honey bees (*Apis mellifera* L.) as potential Varroa mite (*Varroa destructor*) controls. *J. Apicul. Res. Bee World*, 48, 256–262.

Simone M., Evans J., Spivak M. (2009) Resin collection and social immunity in honey bees. *Evolution*, 63, 3016–3022.

Simone-Finstrom M., Spivak M. (2010) Propolis and bee health: the natural history and significance of resin use by honey bees. *Apidologie*, 41, 295–311.

Singh J..N., Venkataraman T.V. (1947) Pseudoscorpions in beehives in India. *Curr. Sci.*, 16,122–124.

Starks P.T., Blackie C.A., Thomas D., Seeley P.T. (2000) Fever in honeybee colonies. *Naturwissenschaften*, 87, 229–231.

Sturtevant A.P., Revell I.L. (1953) Reduction of *Bacillus larvae* spores in liquid food of honey bees by action of honey stopper. *J. Econom. Entomol.*, 46(5) 855–860.

Subbiah M.S., Mahadevan V., Janakiraman R. (1957) A note on the occurrence of an arachnid - *Ellingsenius indicus* Chamberlin – infesting bee hives in south India. *The Ind. J. Veter. Sci. Animal Husband.*, 27, 155–156.

Sudarsanam D., Murthy V.A. (1990) Phoretic association of the pseudoscorpion *Ellingsenius indicus* with *Apis cerana indica*. In: Veeresh G.K., Mallik B., Viraktamath C.A. (Eds.) Social Insects and the Environment. *Ellingsenius indicus Proc. 11th Intern. Congr. of IUSSI.* Oxford & IBH Publ. Co. New Delhi, India, pp. 721–722.

Thapa R., Wongsiri S., Lee M.L., Choi Y.-S. (2013) Predatory behaviour of pseudoscorpions (*Ellingsenius indicus*) associated with Himalayan *Apis cerana. J. Apicul. Res.*, 52, 219–226.

Tian B., Fadhil N.H., Powell J.E., Kwong W.K., Moran N.A. (2012) Long-Term Exposure to Antibiotics Has Caused Accumulation of Resistance Determinants in the Gut Microbiota of Honeybees. mBio 3, DOI:10.1128/mBio.00377-12.

van Toor R.F., Thompson S.E., Gibson D.M., Smith G.R. (2016) Ingestion of *Varroa destructor* by pseudoscorpions in honey bee hives confirmed by PCR analysis. *J. Apicul. Res.*, 54, 555–562.

Topolska G., Gajda A., Hartwig A. (2008) Polish honey bee colony-loss during the winter of 2007/2008. *J. Apucul. Sci.* 52, 95–104.

Valdovinos-Flores C., Gaspar-Ramirez O., Heras-Ramirez M., et al. (2016) Boron and coumaphos residues in hive materials following treatments for the control of *Aethina tumida* Murray. *PLOS ONE*, 11(4): e0153551.

Visscher P. (1980) Adaptation of honey bees (*Apis mellifera* L.) to problems of nest hygiene. *Sociobiology*, 5, 249–260.

Visscher P. (1983) The honey bee way of death: necrophoric behaviour in *Apis mellifera* colonies. *Anim. Behav.*, 31, 1070–1076.

Wilson M.B., Pawlus A., Brinkman D., et al. (2017) 3-Acyl dihydroflavonols from poplar resen collected by honey bees are active against the bee pathogens *Paenibacillus larvae* and *Ascosphera apis. J. Invertebr. Pathol.*, 124, 44–50.

Woyke J. (2001) *Varroa destuctor* (Anderson i Trueman 2000). *Beekeeping*, № 1, 9 (in Polish).

Żyłowska M., Wyszyńska A., Jagusztyn-Krynicka E.K. (2011) Defensins – peptides with antibacterial activity. *Post. Mikrobiol.*, 50, 223–234 (in Polish).

www.ingramcontent.com/pod-product-compliance
Lightning Source LLC
Chambersburg PA
CBHW042226010526
44111CB00046B/2972
* 9 7 8 1 9 1 3 8 1 1 0 2 0 *